W0114684

Praise for
THE CYBERNETIC SOCIETY

"[An] accessible volume...grounded in the belief that cybernetic-AI hybrids could create less bureaucratic, more equitable systems. Whether this future can bridge the digital divide and serve the many remains uncertain—but it is a question worth serious thought...an optimistic, shimmering image of a world where AI operates in service to humankind." —*Kirkus*

"Human beings became the dominant species on planet Earth, surpassing their fellow apes, because they mastered technology. And technology was always a good servant. Now we finally face the prospect of technology dominating humans. Amir Husain explains how and why these new dangers have emerged and how to prepare for them. With his unique multicultural background, he provides new insights we must reflect on. A must-read."

—Kishore Mahbubani, former Singapore Ambassador
and author of *Living the Asian Century*

"In *The Cybernetic Society*, Amir gives an insightful perspective on how we may respond to an increasingly complex world that will inevitably see humans and machines shaping the future together. Highly recommended."

—Peng-Yam Tan, Chief Defense
Scientist of Singapore

"Husain expertly introduces the human and machine decision processes and how they become one—both in the future and arguably already to a large extent today. Should we embrace or fear such a cybernetic society? The book leads us to a fascinating intellectual journey and provides not just some answers but also thought-provoking questions."

—Haibin Xu, General Manager of
Shell Research Alliance

"Amir Husain makes a vivid and convincing case of how AI is shaping our society and lives. The symbiotic relationship between humans and machines called cybernetics will be both an opportunity and challenge. Opportunity because the total human and machine intelligence will be more than their sum, and challenge because it will extend to all aspects of our society, from politics to military to economics. Laced with authoritative references, *The Cybernetic Society* is a lucidly written compelling account of the unfolding AI revolution."

—Pravin Sawhney, author of *The Last War*

THE
CYBERNETIC
SOCIETY

HOW HUMANS AND MACHINES WILL
SHAPE THE FUTURE TOGETHER

AMIR HUSAIN

BASIC BOOKS
New York

Copyright © 2025 by Amir Husain

Cover design by Emmily O'Connor
Cover images © Andriy Onufriyenko/Moment Via Getty Images;
© Imagewell/Shutterstock.com; © Orfeev/Shutterstock.com
Cover copyright © 2025 by Hachette Book Group, Inc.

Hachette Book Group supports the right to free expression and the value of copyright. The purpose of copyright is to encourage writers and artists to produce the creative works that enrich our culture.

The scanning, uploading, and distribution of this book without permission is a theft of the author's intellectual property. If you would like permission to use material from the book (other than for review purposes), please contact permissions@hbgusa.com. Thank you for your support of the author's rights.

Basic Books
Hachette Book Group
1290 Avenue of the Americas, New York, NY 10104
www.basicbooks.com

Printed in the United States of America

First Edition: August 2025

Published by Basic Books, an imprint of Hachette Book Group, Inc. The Basic Books name and logo is a registered trademark of the Hachette Book Group.

The Hachette Speakers Bureau provides a wide range of authors for speaking events. To find out more, go to www.hachettespeakersbureau.com or email HachetteSpeakers@hbgusa.com.

Basic Books copies may be purchased in bulk for business, educational, or promotional use. For more information, please contact your local bookseller or the Hachette Book Group Special Markets Department at special.markets@hbgusa.com.

The publisher is not responsible for websites (or their content) that are not owned by the publisher.

Print book interior design by Bart Dawson.

Library of Congress Control Number: 2024048370

ISBNs: 9781541605718 (hardcover), 9781541605725 (ebook)

LSC-C

Printing 1, 2025

For my sons, my lions, for the next generation, Asas, Murtaza, Hyder, Vali, and Suleyman—

May you always embrace the infinite possibilities of your minds, and may your curiosity, strength, and compassion guide you to shaping a good future for all of us, for all humanity. This book is just one more expression of my love for you—may you hold on to faith, to science, to each other, and may you forever explore the internal and external Universe with courage, humility, and wonder.

With all my love.

CONTENTS

CODE, CONSCIOUSNESS, AND CONTROL

I am a technologist, programmer, and entrepreneur. I've worked with computers and technology my entire life. The AI company I started in 2013 as its sole founder achieved a $1.4B valuation while I was CEO. In 2018 I partnered with one of the largest aerospace companies in the world to launch another business: the first AI company focused on integrating millions of autonomous aircraft into commercial airspace. I've worked with dozens of four- and three-star generals, admirals, and civilian leaders across government to imagine, conceive, and develop ideas and technologies that will shape the future. I've been asked to brief many government leaders in countries across Europe, Asia, and the Middle East, as well as the boards of some of the largest global companies. Finally, I've served on the board of the UT Austin Department of Computer Science, one of the top ten CS schools in the nation. All of this is to say that technology research, implementation, policy, and the future implications of innovation are most of what I have been immersed in for decades. In 2015, before the AI "craze" was mainstream, I began writing *The Sentient Machine*, which explores what we as humans would do in the age of artificial general intelligence. It seemed quite far then, but not so much now. Yet all of this is in the past. The future I see ahead is, at once, more fantastical and yet more likely than anything I have imagined in the past. And that future is what this book is about.

Code, consciousness, and control are three elemental constructs of the future. Because we're awash in devices and websites, we've all seen the power of software. The ability of code to turn a PhD dropout into a multimillionaire is one aspect of this power. Another is its ability to propel a company to the very top of a market, outpacing banks, oil companies, and conglomerates. Code has helped dematerialize things that were once physical and tangible into digital, ethereal constructs. When veteran investor Marc Andreessen talks about "software eating the world," he is simply describing what should by now be evident to us all. Look at an electric vehicle, for instance. It replaces complex mechanical components such as pistons, valves, carburetors, accelerator wires and levers, spark plugs, air filters, oil filters, and much more with a digital control system, electric motors, and a battery. A lot of mechanical complexity and actual physical parts are simply "eaten up." And it's the code that's doing the eating up. And what's happening to the car is happening to the world. Dematerialization. Objects replaced with software.

Code is a fundamental, elemental building block of the modern world. But what can you build with it? Of course, you can build spreadsheets, email applications, and websites. But it turns out that the most remarkable thing you can change, evolve, and perhaps even build with code is consciousness. And I don't just mean this in the sense of AI being a construct of code and AI eventually becoming conscious. I mean this in the sense of human augmentation and the enhancement by digital means of that which is already conscious. Of code embedded in trillions of devices that all come together to influence human consciousness. I mean it in every sense imaginable, and some we can't yet imagine.

And finally, code can be used to control. Of course, we know that even simple devices like thermostats can control a cooling system. A complex web of sensors implemented in a chemical-manufacturing process can control machines, temperatures, flows, and much else on the production floor. But code is also used to control people. It is used to shape and influence thought, shift views, and reprogram minds at a massive scale. Critically, the scale at which it can achieve this is not limited to an individual or a group but is at the level of entire societies.

Are code, consciousness, and control all on equal footing? My answer to this question is personal, colored by my own experiences and, in the absence of an existence proof, my subjective opinion. I believe code to be the most important of these three phenomena. And yes, I call code a phenomenon because it can evolve on its own in a complex system. After all, DNA evolved in the complex crucible of a four-billion-year-old planet, itself a product of a thirteen-billion-year-old universe.

Perhaps one can call such a time-consuming process inefficient; perhaps, looking at many human specimens who are the most advanced manifestation of this evolution, one can even call it imperfect. But a phenomenon it is. Consider, for instance, how the genetic code of a simple single-celled organism has evolved into the complex neural code that governs human cognition.

Thanks to the British mathematician Alan Turing, we now know there is a thing called computational universality. In other words, machines, substrates, and devices of various designs and made from entirely different materials can all be equivalent in their ability to solve any solvable problem. They may differ in how long they take, how large they are, or the resources they consume, but theoretically they are all equivalent. For example, a modern super-computer and a Commodore 64 from 1982 are both capable of performing the same calculations—the difference lies in efficiency and scale, not in fundamental capability.

My corollary, then, is that code can take root in any number of underlying systems capable of performing computation. To the extent that code, when manifested in the world as more than an idea, is an internal organization of the computing system on which it runs, code along with the physical mechanism that executes it can be thought of as a particular organization of matter. This concept bridges the gap between the abstract nature of code and its physical manifestation, much like how our thoughts (abstract) are ultimately the result of neuronal firing patterns (physical).

Any system that runs long enough and can mutate and transmogrify matter is likely to give birth to computational machines. The useful ones are those in which useful code is present as an organization of the underlying matter on which it runs. Everything else is secondary. This principle applies not just to silicon-based computers but potentially to any system capable of information processing—from quantum computers to theoretical biological computers.

If we abstract away all the physical aspects of such systems, better and faster ways to create code mean better and faster results, one of which is consciousness, and another is control. If indeed we can create intelligence that is an existential proof of the emergence of consciousness and control as products of code, it would finally prove the deep, fundamental connection among these three concepts. Just as genetic code gave rise to biological consciousness and neural configurations enable our control over our bodies, perhaps more potent configurations of code will lead to new paradigms of consciousness and control that we can scarcely imagine today.

Be that as it may, in order to understand the world of the future, one must understand code, consciousness, and control. But with what lens should we view all of this? What organizing mental models and principles of integration should we apply to synthesize so much that is happening across finance, technology, military, and foreign affairs to build a holistic picture of the world of the future?

The best answers I have been able to produce are shared with you in this book. The organizing principle we use to tie all these diverse ideas together is cybernetics, the study of control and communication in complex systems.

As you read ahead, you will come across Geoffrey West's scaling laws, which are well-known and studied in biology, in urban planning, and in the context of organizations. But here you will also see these laws applied in the context of mental amplification. You'll read about Peter Turchin's ideas on cliodynamics and elite overproduction in context of the technological shifts they can drive. And you'll journey with me as we tour future technology-enabled metropolises such as Neom and imagine the future of our cybernetic world.

The strangely futuristic sounding field of cybernetics was conceived by MIT professor and polymath Norbert Wiener in the 1930s. Wiener had been born into and lived in a world that was rapidly industrializing. He recognized that automation meant that the relationship between humans and machines would evolve significantly. So, rather than continuing to view humans and machines as separate entities, he proposed that we humans begin to consider ourselves and our machines as a unified whole. Such a perspective has the advantage of encompassing the collaboration between humans and machines, where their behavior and performance become composites of biology, computation, and mechatronics.

In a 1948 paper, Wiener and his colleagues defined *cybernetics* as "the scientific study of control and communication in the animal and the machine." This broad definition allowed cybernetics to encompass fields as diverse as engineering, biology, psychology, and sociology.

Over the ensuing seventy-five years, we've seen waves of enthusiasm for cybernetics, focusing on everything from technology to social and philosophical concerns. Now we find ourselves entering an era in which automation, sensorization, and synthetic intelligence pervade every aspect of the physical world. Cybernetics no longer applies solely to operators and machines but also extends to understanding the future of politics, economics, sociology, and militaries. These emergent systems result from the interaction among humans, machines, code repositories, and synthetic nervous systems. In this new reality, many worry about unemployment and humans displaced by machines, but it may be more beneficial for everyone to understand the nuanced field of cybernetics and how it will likely affect the coming decades.

The scale of this interaction is staggering. The International Data Corporation (IDC) estimates that by 2025, the amount of data generated annually will reach 175 zettabytes. This equates to over 20 terabytes of data per person on Earth every year. We are our decisions and knowledge, but we are also increasingly the data that devices gather on our behalf and the actions that machines execute for us.

Consider the smartphone in your pocket. It's not merely a communication device or a consumer electronic; it's also an extension of your mind, memory, and body. It knows your location, your habits, your social connections, and often your most intimate thoughts.

This cybernetic augmentation of our capabilities is precisely what Wiener envisioned when he coined the term *cybernetics*: the seamless integration of human and machine.

This may sound futuristic, fantastical, or simply unreal. However, even just talking about it makes it real. Reflexivity, as introduced by George Soros in his book *The Alchemy of Finance*, reveals that thinking and behaving as if something is true can make it true. Soros demonstrated this principle in financial markets, showing how investors' perceptions can influence market fundamentals, which in turn reinforce those perceptions. When machines join humans in holding assumptions and acting upon them, the potential for materializing those assumptions becomes a fascinating phenomenon.

The types of autonomous and semiautonomous systems we are now building form beliefs and make assumptions based on observation, and they usually involve an opaque decision-making process through a neural network to execute an action. This leads us to explore whether billions of machines believing in an outcome and acting accordingly can make that outcome a fait accompli. And this can happen in areas as small as an individual athlete or musician's career and also in vast scopes such as the outcome of an election and the choice of specific representatives who make it into Parliament. Today, the visual performance of athletes is already being studied by computer vision systems that break down specific responses, styles of play, and many more aspects of performance for which we humans don't even have a name. Based on all this observation, machines can already provide an idea of which athlete is likely to do better in a particular situation. Picking this athlete over another and even how much to pay them are decisions that are already and will continue to be guided by algorithms. And although I don't know of any mega-donor now who is using artificial intelligence to determine which of a panoply of candidates to back in an election in order to pursue their political and business aims, I would be surprised if this too isn't happening already.

So cybernetic systems are affecting us today, and the quantum of their effect will only continue to increase. Just as it makes sense to understand how your body works so you can make smart choices about your health, it's crucial to understand the technology that will shape us so we can make smart choices about our cybernetic selves. The Apple-ification of technology has made it so that complex, powerful devices seem to be oversimplified appliances. But these systems and the software they run are not mere appliances; they are integral parts of our extended cognitive apparatus. To stay relevant and maintain control over our lives in the cybernetic age, we must push beyond the notion of technology as black-box appliances and strive for a deeper understanding and mastery of these tools that have become extensions of ourselves. If we don't, well, then the machines will likely imagine a future for us and make it real.

The concept of reflexivity in financial markets has been empirically studied by many, including Zhong and associates of the Zhejiang University of Finance and Economics in Hangzhou, China. Their research, published in *Physica A: Statistical Mechanics and Its Applications*, suggests that reflexive feedback loops between market prices and underlying economic fundamentals can indeed lead to self-fulfilling prophecies. By incorporating market impact and momentum traders into an agent-based model, they investigate the conditions for the occurrence of self-reinforcing feedback loops and the coevolutionary mechanism of prices and strategies.

Their study found that when individual trades don't significantly affect market prices (low market impact), traders who follow market trends don't cause large price swings. However, they disturb the balance between those who follow trends and those who go against them. This leads to more people adopting trend-following strategies, creating a self-reinforcing cycle where trends become stronger simply because more people are following them. On the other hand, when individual trades have a big impact on prices (high market impact), these trend-following traders cause larger price fluctuations. In this scenario, smart traders start to avoid following the trends, leading to a negative feedback loop in which trend following becomes less attractive.

These findings underscore how the behavior of traders in financial markets can amplify trends and create feedback loops that influence market outcomes. This kind of self-reinforcing or self-correcting behavior doesn't just happen in finance; we can see similar patterns in many other areas of life as well. For example, LLMs (large language models) are widely used today to write news articles. The choice of words they use to fill in details or provide explanations is critical. Is the organization Hamas a group of freedom fighters? Are they Palestinian separatists? Are they a "proxy group"? Are they terrorists? Whatever the LLM believes makes it into an article that, when read by millions, transforms a machine belief into a human belief—or, at the very least, strongly influences human beliefs.

As AI systems become more integrated into decision-making processes across various domains, keeping an eye on all the other areas where these reflexive dynamics emerge will be nothing short of fascinating. The interaction between AI-driven decision-making and market behaviors could lead to new forms of self-reinforcing feedback loops, making this an area ripe for further exploration.

Could the very knowledge that an AI agent can create such feedback loops allow it to evolve strategies that seek to deliberately create such loops? For example, if a machine knows that the bestseller or "Editor's Choice" list it publishes influences purchases, and it is running out of an inventory of the truly popular products, does it "fake" a bestseller to create demand by including an abundantly stocked product on the list? I would hazard a guess that many AI-driven systems seeking to optimize a goal such as yield or profit do similar things already. And if it happens in trading and e-commerce, can it also happen in politics? In fashion? In any area of society where trend followers and trend rejectors share the psychology of the traders Zhong and colleagues studied? Could algorithms then discover ways in which a goal can be achieved if they believe the goal can be achieved by creating the right feedback loops?

Imagine that. A reality that emerges because machines believe in an outcome. If this is about to be so, then a new digital age now dawns with consequences so deep and profound that nothing we have read or experienced could have prepared us for what is to come.

Many commentators are worried about how many jobs AI might eliminate and whether new jobs will materialize. Already we've seen live customer support replaced with chatbots. We've seen content writers replaced by generative AI. We've seen fast-food chains install robotic kiosks. Our smoothies and drinks are now being delivered by robots. I recently even had a robot live-manufacture highly customized fragrances for my mother and my wife at Dubai's fantastic Museum of the Future. But the transition we are undergoing will have consequences far more profound than changes in the statistics around human employment. We are moving from a world of some physical mystery into a pervasive, sensorized landscape—a reality that is predicted, analyzed, computed, and most likely influenced by large-scale socio-technological cybernetic systems. The implications of this paradigm shift will reshape not just the way we perceive our world but also how we interact with it and, fundamentally, who holds power within it.

A 2019 study by the Brookings Institution estimated that 36 million American workers, or about 25 percent of the US workforce, face high exposure to automation in the coming decades. However, the same study also highlighted that this technological shift will likely create new job opportunities, particularly in sectors that require uniquely human skills such as creativity, empathy, and complex problem-solving. What effect will be dominant? It is too early to tell.

The rise of artificial intelligence serves as the catalyst for this transformation. Deep learning, the most successful category of AI, relies heavily on extensive data to understand, model, and predict various processes, individuals, or systems. The increased demand for data accelerated the proliferation of sensors, which are now ubiquitous, powerful, interoperable, and growing exponentially. However, our understanding of living in a sensorized world has not kept pace with this technology, and we must not overlook the societal shifts that these advancements bring.

If you project the IDC prediction on data creation that I shared earlier, the global datasphere will expand from 45 zettabytes in 2019 to 175 zettabytes by 2025. This means that the amount of data being produced in 2025 will be nearly four times the total amount of data that existed in 2019. The datasphere is like a red giant star, fast encompassing the erstwhile analog solar system within the embrace of its digital glare. In fact, one of the reasons I think reflexivity applies to machine recommendations is that humans can't deal with the quantities of data we have now produced. Humans lose glucose and tire fast when they are forced to make considered decisions. They opt for default behaviors. Quite soon, the default behavior might simply be to click "accept" on the machine recommendation. And if clicking on "accept" signals to the machine that it did well, then such a "reinforcement learning from human feedback" (RLHF) signal could compel the machine to present the highest-value decisions just when the human is running low on decision-making glucose: before lunch or close to the end of the day.

One noticeable implication of deep sensorization and "code as control" lies in the revolution in military systems, where AI renders traditional advantages obsolete. Autonomous drones, cyber-warfare, real-time surveillance, and predictive analytics represent only a fraction of the potential developments. Power projection is no longer limited to those with physical resources but extends to those who can effectively harness AI and the sensorized world. This includes the so-called middle powers, countries with smaller populations but larger budgets, and even corporations. Elections can be influenced with technology. For those who know how to maximize the efficacy of this technology, the cost may not be prohibitive. Large language models shaping political opinion. Deep-fake videos convincing voters. Or convincing the enemy's civilian population that their military has lost an ongoing campaign. Or even that their generals are involved in corruption. How much money would such campaigns take with today's technology? Not much. At another level, if we know that a company or country makes decisions based on sensors that feed an autonomous or semiautonomous system, then we might come up with approaches to influence those sensors. What happens when seismic surveys looking for oil in a contested part of a country's economic zone are deceived by autonomous systems that jam and confuse the acoustic and vibration sensors being used? To my knowledge, this has never happened thus far. But it is just one example of the high-level mayhem that would be caused if we continue to live on Earth under the assumption that we are in a pre-cybernetic age where machine decision-making and its influence on human beings can remain an afterthought.

The US Department of Defense's 2021 budget request included $841 million for AI-related research and development, a 14 percent increase from the previous year. But with the upcoming contract award for a new autonomous unmanned fighter aircraft, this number will jump into the billions. These investments reflect the growing recognition of AI's potential to revolutionize military capabilities, from enhancing situational awareness and decision-making to enabling autonomous systems and predictive maintenance.

The blurring of traditional boundaries necessitates an examination of the interactions between humans and machines, the impact of the sensorized world on social structures, and the resulting feedback loop.

When I was seven or eight years old and enrolled in an all-boys elementary school, I remember how important it was to each of us to be seen in a positive light by our teachers and, quite importantly, by our sports instructors. That was what determined the social hierarchy at the Junior School at Aitchison College. I particularly recall one ex-military physical-training instructor who would sort and organize boys by skill. We all wanted to be in the high-potential group. Today, companies like Ballogy, based in Austin, Texas, use machine vision and artificial intelligence to recommend talent to talent scouts. The "coolness" hierarchy of my elementary school was all about convincing our tough ex-army coach. Today, it may be influenced by styles and approaches that convince algorithms. The fusion of the biological and the digital, the human and the machine, carries far-reaching implications that are still unraveling.

The most obvious realm of cybernetic connection involves brain–computer interfaces (BCIs). They are advancing rapidly, with the potential to directly connect human brains to digital systems. We—humans and machines—may become one entity, tied surgically to our computational technology. The nonintrusive brain–computer interfaces we might soon develop will be much less painful than a common surgery, such as a knee replacement, is today. In fact, we may soon develop interfaces that could simply be strapped on without any need for surgical implantation. In 2021 a team of researchers from Brown University demonstrated a wireless BCI that enabled a paralyzed individual to type using only their thoughts at a rate of 90 characters per minute, showcasing the potential for seamless human–machine interaction.

The intertwining of human societies and machine systems presents difficult questions but also brings forth unfathomable opportunities. The philosophical, legal, and societal considerations are vast and intricate. How do we navigate a world where machines constantly monitor and analyze our lives? How do we ensure equitable distribution of the benefits in a sensorized world? How can privacy and individual rights be protected when every action is traceable and analyzable? As long as we can answer these questions reasonably well, the future will be bright beyond our wildest imaginings. For, with the unlimited cognitive potential we can apply to every decision, we can generate miraculous outcomes: end disease, produce an abundance of food and energy, and solve freshwater shortages the world over. Indeed, we could be living in a world that outdoes our most utopian ideals. What might such a utopia look like, you ask?

Imagine a future where AI manages distributed-solar-energy systems with unprecedented efficiency. An average US family consumes about 10,950 kWh per year as of 2024, equal to about $3,100 in kWh terms, and about $3,875, including all the utility surcharges, if you're in California. In that same state, you have to earn about $5,626 in pretax annual income in order to pay for an average home-electric bill alone. Across the nation, you're talking 5–10 percent of the per capita income of each state being allocated just to home-electric bills. Today, the average person works slightly over a month a year just to afford electricity.

Now picture AI-driven production methods significantly reducing the cost of solar panels. We've made substantial progress—the price of solar has plummeted from $7.53/watt in 2010 to between $2.39 and $3.66/watt in 2024 for residential systems in the United States. However, in countries like Pakistan, we're seeing even more dramatic price drops, with some solar panels costing as little as $0.16/watt in 2024. This trend isn't just about energy independence; it's also about moving toward energy abundance.

When I last visited Pakistan, in the summer of 2024, I was struck by the sight of solar panels everywhere: atop residential roofs, powering lampposts on the streets and emergency phones on the highways, and running tube wells in agricultural areas. I did a little digging and learned that Pakistan's solar-panel imports have surged dramatically, from 2.8 GW in 2022 to 5 GW in 2023, with projections exceeding 13 GW in 2024. This rapid growth is driven by falling panel prices and increasing demand for renewable energy. In fact, Pakistan has become the third-largest destination for Chinese solar-panel exports.

The impact of this solar boom is significant. Pakistan is expected to spend more than $3.5 billion on solar-panel imports in 2024 alone, not including batteries, inverters, and other auxiliary items. The potential, of course, is unlimited. The World Bank suggests that using just 0.071 percent of Pakistan's area for solar-photovoltaic-power generation would meet the country's entire electricity demand. My hope is that much of this is distributed-power generation in the near term, amplifying the economic benefits for ordinary citizens.

But this robot-fueled solar revolution isn't just happening in Pakistan. Globally, we're seeing increased adoption of solar energy, driven by falling prices and the need for sustainable energy solutions. If you trace the supply chain, it's the robotic production capabilities installed in China that are enabling consumers worldwide to benefit from increasingly affordable solar electricity. The rapid advancements in solar technology and its increasing affordability are paving the way for a cleaner, more sustainable energy future. And AI and robots are helping us do it.

Now let's talk about food. The average American spends about 11.2 percent of their disposable income on food. That's almost six weeks of labor annually just in order to eat. As agricultural processes evolve with AI and robotics, we're approaching a point where AI-powered sensors and robots can become competent enough to manage food production inside controlled environments. These systems, powered by expanding solar-energy capacity, could autonomously produce food and harvest seeds. Although we're not quite at production-to-plate yet, we're seeing significant advancements in vertical farming and precision agriculture that are changing our relationship with food production. Less than 2 percent of the US population is involved in producing food today. Perhaps at some point, this will begin to approach zero, and the benefits will be that food security is assured to all as a fundamental right.

What about travel and the need to move from place to place? Well, first, virtual reality (VR) and augmented reality (AR), coupled with AI-based characters and generative voice synthesis, are making strides in blurring the lines between physical and digital existence. Yes, we're not quite controlling our remote robot bodies through AR and a 6G cell network yet, but these technologies are opening up new ways to experience and interact with our world. In fact, Apptronik, a humanoid-robotics company my wife and I are investors in, along with Figure.ai, a competitor in the same space, already allow humans to train robots using AR.

Other than needing to eat and heat or cool our homes, we also need to get from one place to another. This is another area where a utopian future of environmental soundness, low cost, and high convenience is quite possible. Electric vertical takeoff and landing (eVTOL) aircraft are progressing, with several companies conducting successful test flights. My good friend Brian Yutko, the CEO of Boeing-owned Wisk, and my successor CEO at SkyGrid, Jia Xu, are working hard to make it happen. We haven't reached the point of democratized air travel or rendering pilot licenses obsolete, but the potential for revolutionizing personal transportation is very much on the horizon. Companies like Wisk, Joby Aviation, and Vertical Aerospace collectively have billions of dollars of capital on hand to build such aircraft. If regulators, such as the FAA, collaborate with this emerging sector, we could see autonomous vertical-takeoff aircraft ferrying passengers and cargo all over our metropolises within a decade.

In health care, AI-assisted robotic surgery is already a reality. The da Vinci Surgical System, which has been used in more than seven million minimally invasive procedures worldwide, employs advanced robotics and computer vision to enable surgeons to operate with enhanced precision, dexterity, and control. As these systems continue to evolve, they have the potential to democratize access to high-quality surgical care and to improve patient outcomes. I've watched in awe as scientists at UT Austin's Anna Hiss Robotics Lab have demonstrated to me the new algorithms they are developing for da Vinci. Watching a live demo about the level of automation that can soon come to surgery is a mind-warping experience. If robots and AI can provide these interventions at scale, perhaps the cost of medical care can also be reduced in our imagined utopian future.

But none of this matters if we live on an environmentally stressed planet. Earth is the only home we know and the one that any future of technology must work to preserve. AI is increasingly being applied to environmental challenges. Although we're not yet at the stage of AI-synthesized materials actively healing our planet,

AI is playing a crucial role in climate modeling, resource managing, and developing more efficient carbon-capture technologies.

This may sound like science fiction, but remember that the future has a way of sneaking up on us. Cybernetic You 1.0 already exists. That's human you augmented by your smartphone, today's generative AI, and all the apps you use. Cybernetic You 1.0 already has greater technological sophistication and access than even President Bill Clinton could muster during his tenure. And Cybernetic You 2.0 is closer than you might think. The systems we're building today—next-generation AI models; more powerful AI processors to run intelligence in small, portable devices; neuromorphic processors inspired by the human brain; and very low-latency, high-bandwidth networks—are all laying the foundation for the next version of you.

My fear, of course, is that this happy future isn't inevitable. It's ultimately a choice. The work needs to be done. The policies need to be adopted. The vested interests have to be fought. Will we use our cybernetic potential to create abundance or to consolidate power? Will we extend our technological capabilities to experience more of the world or to retreat from it? Will we heal our planet or continue to exploit it? The answers to these questions aren't predetermined. They depend on us—on our understanding of these technologies, on our ability to guide their development, and on our willingness to embrace a future that might seem uncomfortably different from our present.

There does, of course, remain the risk of actively doing bad things with this embarrassment of technological riches. A 2020 report by the World Economic Forum identified the responsible use of technology as one of the key ethical challenges in the fourth industrial revolution. The report emphasized the need for inclusive and transparent governance frameworks to ensure that the benefits of emerging technologies are distributed equitably and that potential risks are mitigated proactively.

The UN Secretary-General's High-Level Panel on Digital Cooperation, in its 2019 report "The Age of Digital Interdependence," called for a multi-stakeholder approach to address the challenges posed by the rapid development of AI, including its military applications. The panel recommended the establishment of a Global Commitment on Digital Trust and Security to promote the responsible development and use of digital technologies. Yet we live in the times of the great tech divide between the United States and China—just when intelligent people around the world clearly recognize the need for more collaboration.

Then there's my fear of mass fear. A 2020 survey by the Pew Research Center found that 52 percent of US adults are more concerned than excited about the use of artificial intelligence in daily life. Engaging the public in an open and transparent dialogue about the development and deployment of AI systems, particularly in sensitive domains such as military applications, will be crucial for building trust and support.

Understanding these complexities is essential, and proactive measures must be taken to address ethical concerns, operational uncertainties, concerns of livelihood, and geo-strategic implications. These responsibilities cannot lie only with military commanders but must also span across the entire life cycle of each system, including with citizens, politicians, regulators, and developers. Preparing to handle this new reality requires organization, training, and equipment. And although these transformations are almost guaranteed, citizens must be taken onboard. Public outreach is necessary to inform stakeholders of the military's commitment to mitigating ethical risks and avoiding potential policy limitations and public backlash.

I hope to deepen your understanding of this new landscape and equip readers with the knowledge to navigate it confidently. As we embark on this journey together, I invite you to explore the cybernetic-sociological perspective of our increasingly sensorized world. In the pages ahead we will delve further into these concepts, exploring their implications and offering strategies for navigating this new terrain. I believe that this book can serve as a compass that might guide you through the intricacies of our sensorized world, exposing its risks, highlighting opportunities, and illuminating the path ahead. As we venture into uncharted territory, I encourage you to suspend disbelief about the future, question the status quo, and join me in shaping the necessarily complex but absolutely fascinating cybernetic world of tomorrow.

CHAPTER 1

THE ORIGINS OF CYBERNETICS

In the summer of 1947, a brilliant, enigmatic, fifty-three-year-old man with a deeply thoughtful face marked by sharp, penetrating eyes stood at the forefront of a scientific revolution. This man was Norbert Wiener, an American mathematician and philosopher. Together with Arturo Rosenblueth, a talented physician born in Mexico, Wiener saw machines as no one prior to him had quite seen them. He conceived of them not as independent systems but rather as fused with their human operators—parts of a whole. This idea of human–machine symbiosis birthed a new field of study: cybernetics.

Seen from the vantage point of today, this was indeed a pivotal moment. The birth of cybernetics and its evolution over the past century have silently built up the groundwork for the profound changes we are witnessing in the interplay among code, consciousness, and control. We can now begin to see how the early ideas of pioneers like Norbert Wiener and André-Marie Ampère have set the stage for the emergence of complex, interconnected systems that blur the lines between human and machine cognition, and even combine the two into a Gordian Knot of thought impossible to unravel. The concept of feedback loops and the interplay between biological and mechanical systems, central to the cybernetic paradigm, is only now manifesting as a form of "fused" control via AI-driven decision-making, autonomous systems, and the increasing integration of technology into every aspect of our lives.

Wiener and Rosenblueth's investigation into the phenomenon of control in biological systems and machines would go on to substantially influence our understanding of complex systems. Wiener was part of a chain of great minds obsessed with the science of control and the nature of feedback mechanisms. The term *cybernetics*, derived from the Greek word κυβερνήτης (kybernētēs), meaning "those who steer or govern," has historical roots dating back to the nineteenth century.

More than a century earlier, in 1775, André-Marie Ampère, a French physicist born in Lyon, had gained worldwide fame for establishing the science we now call electromagnetism. His name endures in everyday life in the ampere, the unit for measuring electric current. Ampere developed a mathematical and physical theory to understand the relationship between electricity and magnetism. Building on Danish physicist Hans Christian Ørsted's work, Ampère showed that two parallel wires carrying electric currents repel or attract each other, depending on the relative direction of the currents. Ørsted discovered that electric currents create magnetic fields, thus establishing the connection between electricity and magnetism. Ampère went a step further. He found that the mutual action of two lengths of current-carrying wire is proportional to their lengths and the intensities of their currents— a statement known now as Ampere's Law.

Ampère also theorized the existence of an "electrodynamic molecule," a precursor to the concept of the electron, which he proposed as the fundamental element of electricity and magnetism. His 1827 publication, "Mémoire sur la théorie mathématique des phénomènes électrodynamiques uniquement déduite de l'experience," is considered the founding treatise of electrodynamics. But what was to most inspire Wiener's later contributions was Ampère's vision for a science of "governing," for which he used the French word *cybernétique*. This precursor to cybernetics, "the art of governing" or "the science of government," was an early conceptualization of the field that would later be formally built upon and established by Wiener and Rosenblueth.

This early definition by Ampère hinted at the broad applicability of an underlying theory that could be applied across various domains, including governance, resonating with Wiener's conceptualization of cybernetics as a field concerned with systems, control, and feedback mechanisms. Building on Ampère's ideas and related scientific threads that stretched over the past century, Wiener significantly advanced these ideas by formally establishing a new field, giving it a name, and expanding its scope to include the study of control and communication in both animals and machines. His work focused on uncovering an underlying theory of control in systems made up of human and machine. This core idea, that there can be a theory that explains the emergent outcomes of a human–machine system, has deep transdisciplinary implications. It invites us to revisit cybernetics and its relevance in the world of today.

FROM PRODIGY TO PIONEER:
WIENER'S EARLY LIFE

Born in Columbia, Missouri, to Jewish immigrants from Lithuania and Germany, Norbert Wiener was a true child prodigy, displaying remarkable intellectual abilities from an early age. His father, Leo, a linguist and mathematician, had provided him with a rigorous home education that shaped his analytical skills. When he was only ten, Wiener authored "The Theory of Ignorance," a work that questioned the concept of limitless human knowledge. Demonstrating a mixture of precociousness and a rather prodigious intellect, he earned a BA in mathematics from Tufts College at fourteen and completed his PhD from Harvard at nineteen, specializing in mathematical logic. Wiener's academic journey was marked by mentorships under philosophers such as Bertrand Russell and mathematicians such as G. H. Hardy. In working with these giants, he developed an intellectual rigor and a profound understanding of the complexities of both human and mechanical systems. A polymath, Wiener demonstrated diverse interests. He took a stab at journalism, then later accepted a role as an engineer at General Electric. But as the Great War descended upon Europe, Wiener found himself deeply enmeshed in ballistics research. This would lay the groundwork for his ongoing work with the defense establishment and the development of his cybernetic ideas.

After an interlude of sanity following the "War to End All Wars," Europe's many internal disputes once again pushed the world to the brink. With the invasion of Poland, World War II began. Hitler, the German chancellor, and his Oberkommando der Wehrmacht had developed a new strategy called "blitzkrieg," or lightning war. Characterized by a high degree of speed, it was an extreme form of combined-arms-maneuver warfare in which armored columns and attack aircraft moved together, coordinated by radio. German aircraft such as the Stuka dive bomber and Messerschmitt fighters struck fear in the hearts of all opposed. In order to stop the German juggernaut, Hitler's Luftwaffe had to be stopped first, and this was precisely the job Wiener was assigned.

Wiener, then teaching at MIT, began developing a theory of operation that would allow for the automatic aiming and firing of antiaircraft guns. In studying this problem, Wiener combined information theory, computer science, control theory, robotics, human–machine interfaces, and practical automation. This cross-disciplinary work, woven together with the core ideas of cybernetics, would later help shape cognitive science and AI research at MIT.

But where Wiener's theories were deep, the practical results of his work were not. Nor was he gaining a reputation for being particularly easy to work with. Despite his status as a pioneer of the field, he was largely ignored at the now-famous 1956 Dartmouth meeting where Marvin Minsky, John McCarthy, and the fathers of artificial intelligence gathered to discuss how machines could mimic human thought. Wiener came across as self-absorbed and somewhat ornery, prompting the Dartmouth group to distance themselves from the field of cybernetics simply so that they could avoid him. It was obvious that many of the topics these two groups were interested in were quite similar but that their cultures were very different.

The Dartmouth conference marked a pivotal moment in the history of AI. This gathering was where McCarthy pushed to make logical reasoning the center of attention for their study and coined the term *artificial intelligence*, distinguishing it from the broader and more interdisciplinary field of cybernetics proposed by Wiener. Minsky went on to develop the perceptron in 1957, an early form of neural network that influenced machine learning and pattern recognition for decades. The perceptron is the great-grandfather, so to speak, of deep learning. The connectionist approach that perceptrons pioneered saw a great resurgence in the 2010s, bringing AI research and applications to homes and offices everywhere.

In 1958 McCarthy also contributed significantly to AI by developing the Lisp programming language, which became the standard tool for AI research because of its very strong support for symbolic reasoning. McCarthy's work extended beyond programming languages to fundamental theories of AI, emphasizing logical reasoning and problem-solving.

The Dartmouth conference was profound in another sense, in that it proved to be a wellspring of research. It led to the establishment of major AI research institutions such as the MIT AI Lab, the Stanford AI Lab, and the AI research group at Carnegie Mellon University. Each one of these institutions has been instrumental in advancing the field and producing many of the leading researchers and innovations in AI.

Given this history, we can now see that just because artificial intelligence doesn't sound anything like cybernetics, this doesn't mean the two aren't related. In a strange twist of fate, we might be entering a world where humans and AI have already become components of a planet-scale cybernetic environment.

Wiener's idea of cybernetics emphasized the integration of human and machine components within systems and aimed to explain this symbiosis through mathematical models. His theory put forth a view of the world in which feedback mechanisms from both humans and machines operate cyclically, giving rise to system-level behaviors. Neither the machine nor the human alone, but both in unison, make the system.

Can Wiener's cybernetics inspire a new way of thinking about interactions and relationships in society? By viewing social systems through a cybernetic lens, can we understand societal dynamics as complex interplays of communication and control, similar to interactions in technological systems?

Our world today is filled with sensors, AI, and interconnected systems that, together, present large-scale cybernetic constructs. Cameras identify objects and their behaviors, and humans decide what to do about them. Sometimes, human decisions are expressed in the form of machine actions, such as the opening or closing of a gate, the approval of a financial application, or adjustments made to industrial equipment. In other words, these human–machine systems exhibit macro behaviors that influence and are influenced by other cybernetic constructs, suggesting a planet increasingly characterized by cybernetic behaviors. This perspective can provide us with a unique lens through which to analyze and understand the complex interplay of technology, society, and our environment.

CYBERNETIC FUSION

If we observe the world and are conscious of the dynamics of cybernetics, it is clear how systems and human decision-making processes intertwine into complex, inseparable wholes. Once we experience that critical "red pill" moment, as Neo did in the cult classic *The Matrix*, our eyes and mind begin to process reality very differently. Everything from traffic flow to global finance appears to be part of a vast network of interconnected systems. Human actions and decisions become inputs for these systems, influencing and being influenced by a myriad of other factors in continuous feedback loops. This cybernetic perspective reveals a world where the boundaries between human agency and automated systems blur, creating a dynamic environment for decision-making.

THE NATURE OF HUMAN DECISION-MAKING

The Nobel laureate Daniel Kahneman explained the two principal forms of thinking in which we, as humans, engage. Type 1 thinking, or intuitive reasoning, is fast, automatic, and often unconscious. It involves making quick judgments based on heuristics and is influenced by emotions and biases. Type 2 thinking is analytical, deliberate, and slower. It requires conscious effort and is used in complex problem-solving and decision-making situations.

Kahneman pointed out that much of the time, we make decisions based on intuition and impulsive Type 1 thinking. As biological organs, limited in their capacity to produce energy, our brains have evolved to conserve as much power as they can by preventing laborious computation when simple pattern matching will suffice. When we engage in Type 2 thinking, we do so at great cost. We begin to wear down our reserves of glucose, quickly degrading our ability to reason and apply methods of science and logic.

Kahneman's work in behavioral economics, particularly his collaboration with Amos Tversky, highlighted how cognitive biases and heuristics often lead to systematic errors in judgment. For example, the "availability heuristic" causes people to overestimate the likelihood of events that are more readily recalled from memory, often because they are dramatic or recent.

Kahneman, after a tremendously successful career, passed away in Switzerland in March 2024 at the ripe age of ninety.

THE NATURE OF MACHINE DECISION-MAKING

In contrast, machine decision-making is multifaceted and subject to very different considerations. The simplest forms of decision-making in machines are explicit if-then-else statements based on predefined rules and conditions. Here, a machine generally measures an external variable or sensor, compares its value to a predefined threshold, and then engages an action when the threshold is met. Everything from a simple timer that triggers an alarm, to most form validation, to game and application logic implements this type of decision-making.

Expert systems present another type of machine decision-making. Here, software is given access to a range of facts that are then processed with the rules of logic, making deductions that can give rise to yet new facts. All men are mortal. Socrates is a man. From these two facts, the system might deduce a new fact that "Socrates is mortal." Expert systems generally rely on databases of expert knowledge and then use logically derived conclusions to make decisions. These systems are usually focused on a single, specific field. Not sure why your stomach hurts? An expert system can be used to narrow down the possibilities based on your symptoms, helping figure out if it's an ulcer or the extra helping of ice cream you had after the spaghetti dinner. Your minivan is making a funny, clunking noise? Again, an expert system can help figure out if it's the alternator or the timing belt.

Expert systems yielded many successes in the 1980s and 1990s but fell out of favor because of their inherent limitations, their inability to incorporate new knowledge and learn on their own, and the high cost and complexity of interviewing a large enough number of experts to create domain-focused databases of facts.

If expert systems rely on logic, then surely machines can also rely on statistical reasoning to think and conclude. A large number of very successful decision-making software applications use statistical methods to infer probabilities and make predictions based on data. For example, predicting the number of workers who will show up to work tomorrow in an automotive factory is a problem well mapped to statistical techniques. Even simulations that project many futures and calculate most-likely futures, using techniques like Monte Carlo, are statistical-reasoning systems.

One of the notable applications of statistical reasoning is in predictive maintenance. Companies like General Electric and

SparkCognition, the Austin-based company I founded, use machine-learning algorithms to predict when industrial equipment is likely to fail, allowing for maintenance to be scheduled just in time to prevent breakdowns, thus saving costs and minimizing downtime.

Until the early 2000s, lots of enabling work was done and key advances were made that remained relegated to the academic realm, but in 2010, as automation became cheaper and better, AI entered a renaissance. And this renaissance was triggered by large neural-network systems composed of many layers. So many layers and neurons, in fact, that networks that made use of them were collectively referred to as deep-learning systems. Of course, neural networks themselves have been around since the 1940s, but computers simply haven't been fast enough or equipped with sufficient memory to act as a viable computational substrate for truly large networks . . . until now. Everything from ChatGPT to Google's game-playing AlphaGo and protein-decoding Alpha-Fold is powered by neural networks. ChatGPT allows us to enter a simple writing prompt and have the machine compose an essay on the Reconstruction Era in moments. AlphaGo is able to beat even masters at the game of Go without even working up a sweat. AlphaFold may help researchers find new malaria vaccines or break down plastics in our environment. Neural networks are behind the scenes, working to make the previously-thought-to-be impossible possible. They are everywhere, and they are growing.

The success of deep learning eventually materialized because of the availability of large datasets and significant improvements in computational power, particularly the use of graphics processing units (GPUs). Companies like Nvidia have been massive drivers of this phenomenon and have been richly rewarded by investors. As of July 2024, Nvidia's market capitalization stood at a staggering $3.15 trillion.

One tremendous capability of neural networks is how well they can continue to learn from new data by being "retrained." These networks are impressively effective across a very broad domain of tasks: everything from perception to decision-making to control and action. Neural networks are used to process pictures and video, and identify details of objects and scenes. They are also used to control how a robot walks and how a self-driving car drives itself. These algorithms draw inspiration from human brains and, like them, are often opaque.

For example, Tesla's autopilot system uses neural networks to interpret visual data from cameras and make driving decisions. This system also learns from new data collected from Tesla vehicles worldwide, improving its performance and safety over time.

No matter which of these techniques a machine uses—individually, or even in concert with others—machines don't suffer from the need to minimize energy consumption as human brains must. They can keep thinking deeply about issues for hours and hours with no tiredness or loss of concentration. In fact, many new applications based on large language models (LLMs) can spawn individual, specialized AIs that can be a product manager, a developer, a tester, and a CEO, and then enable direct dialogue among these AIs for hours at a time, with no pause or rest or coffee breaks, to develop an application, write a report, or conduct an experiment. We humans would struggle to hold our concentration for such long stretches of time. Not to mention, these AIs would "argue" civilly, with no hurt feelings when their "opinions" are not chosen. Office dynamics could become much quieter and less confrontational, though, one might argue, not nearly as entertaining.

The concept of specialized AIs working collaboratively is becoming a reality in various fields. For instance, IBM's Watson, which famously won on the quiz show *Jeopardy!*, has evolved to assist in medical diagnostics, legal research, and financial planning by leveraging multiple specialized algorithms.

THE SYNTHESIS OF HUMAN-MACHINE DECISION-MAKING IN A CYBERNETIC SYSTEM

In a cybernetic system, these diverse thinking approaches coalesce. Humans provide intuitive and analytical insights, whereas AI systems range from simple rule-based engines to complex, inscrutable neural networks. These systems function through a hierarchy of decisions, from basic automation to complex, data-driven inferences, forming an intricate tapestry of human-machine interaction.

On a macro scale, cybernetic thinking reveals a world where multiple systems interact in complex, often unpredictable ways. In stock markets, for example, each trade is a decision made for a specific reason, yet the aggregate effect on a stock's price can be wildly unpredictable. This unpredictability escalates when considering the vast number of cybernetic systems that interact globally—from automated trading algorithms to social media dynamics—creating a labyrinth of interdependencies. The execution of each trade is reflexive, affecting the stock price in turn. An algorithmic prediction—a machine's belief—that price will go down results in human traders making lots of sell trades, which in turn actually causes the stock price to go down. The machine's prediction has changed the future.

High-frequency trading (HFT) exemplifies this interaction. Algorithms execute thousands of trades per second, reacting to market data and often influencing price movements. This has led to phenomena like the "flash crash" of 2010, where rapid selling by algorithms caused a temporary but drastic drop in stock prices.

In such a world, traditional decision theory must evolve to encompass both statistical and inexplicable elements. The challenge becomes harnessing these systems for better outcomes while grappling with the reality of often opaque decision-making processes. This raises questions about accountability, ethics, and our understanding of cause and effect.

Ethical considerations in AI are paramount, as seen in debates over autonomous vehicles making life-and-death decisions. The "trolley problem" scenario, where a vehicle must choose between two harmful outcomes, exemplifies the ethical complexities in AI decision-making.

It is almost certain that we will find ourselves navigating a landscape where decisions are made on grounds that are not fully comprehensible—because they have been made by layers and layers of AI systems, each of which is not fully explainable—yet have a profound impact on our lives.

EXPLAINABILITY AND POST FACTO RATIONALIZATION

Are neural networks—AI brains—opaque and human decisions transparent? Are all human decisions well-reasoned in advance of the decision being made? How does machine reasoning compare?

The concept of post facto rationalization in human decision-making is quite astounding and shocked me when I first read about it in depth. Human decisions are often contrasted with the lack of explainability of neural-network decisions in artificial intelligence. Yet both phenomena reveal challenges and significant complexity in understanding the "why" behind decisions made by humans and machines.

Humans are often perceived as rational beings by other humans. We assume that most people make decisions based on logical reasoning and objective analysis. However, studies by Kahneman and Tversky in the 1970s revealed the existence of cognitive biases, demonstrating convincingly that humans systematically make choices that defy clear logic. Paradoxically, this irrationality proves beneficial in decision-making because it allows for quicker decisions in complex, uncertain situations without the need for exhaustive analysis of every possible outcome.

Because we so often jump to unreasoned conclusions, in order to continue to maintain the facade of being rational animals—even to ourselves—post facto rationalization in humans has emerged as a mechanism to justify past choices, even those proven suboptimal in hindsight. This is a construct of our fascinating and clever brains attempting to align past decisions with current beliefs or knowledge, often distorting reality to avoid the discomfort of admitting a mistake!

Our behaviors are rooted in biases and heuristics, mental shortcuts that simplify decision-making processes. Although these shortcuts can lead to efficient decision-making in uncertain scenarios, they can and do also lead us to fall victim to the sunk-cost fallacy, making choices based on irrecoverable past investments rather than present and future benefits.

Erik Eyster, Shengwu Li, and Sarah Ridout, a team of Harvard and UC Santa Barbara researchers, developed a theory of ex post rationalization that provides a formal description of this behavior. In their publications they propose that individuals attempt to rationalize past choices by adapting their attitudes or beliefs, thus affecting their present and future decisions. Their model predicts behaviors such as the sunk-cost fallacy and provides a framework for understanding how and why individuals justify past choices that were mistakes in hindsight.

In contrast to the introspective rationalizations of humans, neural networks operate through complex interconnections, and they weight adjustments learned from data. These models, especially deep-learning networks, are often criticized for their "black-box" nature. The decision-making process of a neural network is largely opaque, not because it seeks to hide or justify or rationalize its decisions but because the sheer complexity and volume of operations involved make it difficult for humans to trace and understand the reasoning behind any given output. In other words, the weights are transparent, but their meaning to humans is not.

The fundamental difference between human post facto rationalization and neural-network explainability lies in these decision-making processes. Humans, aware of their choices, can reshape their narratives and beliefs to justify past decisions, driven by emotional needs, cognitive biases, and social influences. Neural networks, devoid of consciousness or emotional stakes, derive decisions from mathematical optimizations and pattern recognitions that are not designed to be introspective or justifiable in human terms.

But perhaps this is only a shortcoming of present AI systems. It is quite possible to develop an AI "brain" made up of an actor and a critic; the actor makes the decisions, and the critic provides post facto rationalizations! A basic system of this type can actually be developed with the LLM technology that exists today. So perhaps humans and machines aren't quite so different in how they make decisions and then seek to hide their mistakes.

The deeper issue, of course, is that since the dawn of time we humans have had this flaw, now documented and formally described by scholars. All of history has resulted from a set of human decisions that have often been rationalized after they were made. Just as we may not fully understand or articulate the subconscious influences and heuristics guiding our decisions, the operations within neural networks are not fully transparent or interpretable to their creators or users. In both cases, the "why" behind a decision can be elusive. As the systems of society grow larger and larger, and the number of interconnected human and machine nodes multiply, our future will be built with error-prone building blocks. Yes, we will be presented with issues of trust and reliability, but these are not unique to machines.

Despite these challenges, there is an important distinction in the quest for explainability. For neural networks, the field of explainable AI (XAI) seeks to make the decision-making processes of AI systems more interpretable and understandable to humans. The expectation we have is that an AI will fully rationalize and explain each decision to our satisfaction with an audit trail, a reasoning graph, or something equally specific. In contrast, human rationalization processes are subjects of less precise psychological study, with efforts focused on raising awareness of cognitive biases and promoting more rational, reflective decision-making practices. We tend to be more accepting of subjective explanations when it comes to human decision-making. Perhaps what is needed is a holistic view of cybernetic systems. Progress will depend on the recognition that human and machine decision-making is flawed in different ways; composing these decisions together and "error correcting" them is likely to provide far better outcomes.

ERROR CORRECTION IN SOCIAL-SCALE SYSTEMS

The concept of building reliable systems from unreliable components has been a fundamental driver of progress in computing. Error correction plays a crucial role in everything from communications to disk-storage systems, allowing the creation of robust and resilient systems that can withstand the inherent imperfections of their constituent parts and the harshness of the environments they operate in. By employing sophisticated error-correction techniques, engineers routinely design systems that operate with remarkable reliability in the face of component failures and even abuse.

One of the most common uses of error correction in computer systems is the use of error-correcting codes (ECCs) in memory and data storage. ECCs, such as Hamming codes and Reed-Solomon codes, add redundancy to the stored data, allowing the system to detect and correct errors that may occur as a result of hardware faults or external disturbances. The idea is to pack more information than is strictly needed in a way that allows any errors to be both identified and corrected. To think of this in simple terms, imagine sending the message "hello" across a noisy communications channel that can change any one of the letters to a different letter. One way in which to resolve these issues is to send the string "hello" three times. If one of the characters has changed, you simply take the majority vote. This redundancy comes at the cost of increased storage overhead, but it ensures that the integrity of the data is maintained even when individual memory cells or storage units fail. Without ECCs, even a single bit flip could render the entire data useless, compromising the reliability of the system.

Of course, error-correction techniques continue to get more and more sophisticated, and they are now being applied to the underlying technology that can help us build brain-like "neuromorphic" technology. One of the underlying memory devices that can enable these brain-like chips is called resistive random-access memory (RRAM). In a February 2023 study published in the *IEEE Transactions on Electron Devices*, researchers demonstrated a new scheme that could employ error-correcting codes and adaptive programming techniques to achieve a bit error rate (BER) of less than 10^{-6}—that's one in one million—a significant improvement over conventional methods. Today, perhaps we are correcting errors in brain-like chips. But tomorrow, could we use chips to augment our brain and "error correct" memory lapses and other problems?

Moving beyond a single human brain, by embracing the principles of error correction we can probably also design social systems that are more adaptable, fault-tolerant, and capable of self-correction. Famously, democracy is a majority-wins decision to elect a leader under the premise that more than half the population cannot be ill-advised or insane. Of course, questions arise: What were the social-cybernetic constructs and systems that gave rise to the choice of candidates to begin with? What did the computer algorithms at Facebook and X do to promote one candidate over another? What type of attack ads did an AI LLM generate, based on human prompts, to outmaneuver other candidates?

In a world where the actions of individuals and the decisions of algorithms can have far-reaching consequences, the ability to detect and correct "errors" becomes paramount. By introducing redundancy, using technology we own and control ourselves to look for anomalies at all levels and increasing diversity in our social structures, we can ensure that the system as a whole remains stable and functional, even when individual components falter or fail: a kind of social-error correction.

The theoretical impact of applying error correction to social-scale systems is profound. It suggests that we can design societies that are more resilient to the challenges posed by human fallibility, technological glitches, and environmental disturbances. These benefits could manifest in the form of more robust governance structures, more adaptive economic models, and more inclusive social policies that can withstand the test of time.

So, if you stop to think about it, the application of error correction in cybernetic constructs may have implications that extend beyond the realm of technology. As we integrate more deeply with machines and algorithms, the concept of error correction is not just something that can make our home broadband more reliable; it can also make our institutions and governments finally do what they should have been doing all along: perform well in our interest.

CHAPTER 2

NEOM AND
THE WORLD OF
THE FUTURE

In the sun-drenched sands of the Arabian Desert, by the banks of the Red Sea, a project unlike any other is taking shape—a bold vision to create not just a city but also an entirely new model of urban living. Neom is the brainchild of Prince Mohammad bin Salman and part of Saudi Arabia's Vision 2030. It represents the most ambitious attempt yet to bring the principles of cybernetic urbanism to life at a massive scale.

Whether or not it materializes exactly as it has been envisaged, Neom embodies a radical rethinking of the relationship among humanity, technology, and the built environment. It is a grand experiment in fusing human cognition and ingenuity with the power of artificial intelligence and other advanced technologies, creating a symbiosis within which the boundaries between the physical and digital can dissolve.

Neom is remarkable because it reimagines the city not as a static, inflexible structure but rather as a living, breathing organism—one that adapts and evolves in real time to the needs and desires of its inhabitants. By integrating AI into every aspect of urban life, from infrastructure and transportation to health care and education, its designers hope that Neom will optimize efficiency, sustainability, and human well-being in ways that were once the stuff of science fiction.

The realization of this vision is no small undertaking. Spanning an area larger than many nations, Neom is envisioned as a gargantuan megacity built entirely from scratch on the shores of the Red Sea. With a projected population of nine million citizens and an estimated cost of more than $500 billion, the sheer scale and ambition of the project is eye-watering.

Yet the true audacity of Neom lies not in its size or expense, but in its aspiration to push the boundaries of what a city can be. The vision of Neom as a fully realized cybernetic city exemplifies the potential for code to exert control over large-scale built environments, melding human and algorithmic decisions in powerful new ways. Neom represents the largest effort thus far in creating a meta-system of component physical AI systems: a city made up of layers and layers of interacting bits of autonomous systems interacting with humans and machines. From the outset, this project has been driven by a singular, overarching mission: to create a blueprint for the future of human civilization—one that brings together technological innovation with environmental sustainability and human betterment.

As the world's first fully realized cybernetic city, Neom has the potential to be a laboratory for testing and refining the principles of an entirely new urban paradigm. Every aspect of the city's design and operation will be shaped by a deep integration of AI and other advanced technologies, creating a self-regulating ecosystem that adapts and evolves in response to the needs and behaviors of its inhabitants. At the heart of this vision is the concept of the "cognitive city"—a metropolis where AI serves not just as a tool for optimization and automation but as a foundational component of the urban fabric itself.

Consider, for example, the city's transportation system. Neom will be entirely autonomous, with self-driving mass transit and intelligent traffic-management systems that optimize mobility while minimizing congestion, emissions, and energy consumption. By training on real-time data regarding traffic patterns, weather conditions, and individual travel needs, the city's AI systems will be able to dynamically adjust routes, schedules, and vehicle allocations to ensure efficient and seamless movement of people and goods.

Or take energy and resource management. Neom's infrastructure will be powered entirely by renewable sources, with AI algorithms continuously balancing supply and demand to ensure a reliable and sustainable power supply. Smart grids and intelligent building-management systems will optimize energy and water use, and integrated waste-management systems will enable the widespread adoption of circular economy principles.

At SparkCognition, now Avathon, the industrial AI company I founded in 2013, we have been implementing AI for some of the world's largest renewable-energy companies, and I have seen the potential of such applications firsthand. Optimizing the production and use of renewable energy is a multifaceted problem that requires AI to predict the weather, predict and model the degradation of batteries in which energy is stored, optimize the yaw and direction of wind turbines, identify soiling and problems in solar panels, and use all these predictions not simply to suggest actions but to undertake them as well. Multiply this type of constantly adapting decision-making thousands of fold, to factor in individual subsystems, batteries, panels, turbines, gearboxes, and more, and you might begin to have a sense of how complex, yet useful, such a system can be.

The true power of Neom's cognitive-city paradigm may lie not just in its efficiency and sustainability but also in its ability to enhance and augment human experiences and capabilities: a city that makes human beings more capable and "better" just because they live in it. As it's being planned, the city's systems should be able to anticipate and adapt to the needs and preferences of individual residents, delivering personalized services and experiences that were once merely glimmers in the eyes of technologists. For example, AI systems are being developed that would enhance the experience of visitors and residents by personalizing recommendations for attractions, delivering virtual tour guides on devices of an individual's choice, and even automating routine tourism services like checking in and out of a hotel or obtaining a local cell phone for lower-cost connectivity. Even more impressively, AI models will constantly simulate a digital twin of Neom, forecasting issues with your water, power, and cooling before these issues occur.

What will it feel like to live in Neom? Imagine your daily schedule being seamlessly coordinated by an AI assistant that anticipates your needs and preferences based on your routine, habits, and real-time data. Imagine your commute to work being optimized in real time based on traffic patterns and weather conditions, and your transportation options being suggested based on your individual needs and preferences. Where your home environment—from temperature and lighting to entertainment and security—is continuously adjusted to create a personalized, optimized experience tailored just for you. At some point, millions of these small decisions to make life just a little bit simpler, and just a bit more convenient, blend into the background. The power never fails. The temperature is always managed to your liking. Your meetings and schedules are always optimized. Your path is always guided and illuminated, literally and figuratively. The cognitive city might begin to feel like life how it should be and, eventually, life as usual. Places that can't provide these layers of intelligent convenience might begin to feel arcane and underdeveloped. What do you mean I can't just walk into a hall and experience life in the Jurassic? Technology raises expectations, and in this way, cybernetic cities might even begin to shift demographics simply because they are far more livable, convenient, safe, and friendly.

If the vision of Neom's designers is eventually realized, personalization and customization will extend far beyond individual services and experiences to community-wide ones. For example, public spaces will transform and reconfigure themselves based on use patterns and community needs, seamlessly adapting to serve as parks, plazas, or event venues as required. This reconfiguration will be achieved with modular design elements and smart infrastructure. Neom planners are incorporating AI-assisted urban planning and design into their development process. By analyzing data on human behavior, the environment, and resource flows, they are creating optimized layouts and infrastructures tailored to the unique needs of each neighborhood. This data-driven approach will inform decisions on everything from building placement to transportation networks; in other words, how the machines and the algorithms they run predict urban resource use is how the city will be built. The machines suggest, and we humans build accordingly.

Although the vision of fully autonomous, self-reconfiguring spaces remains aspirational until we see it play out in real life, Neom is certainly laying the groundwork for such a highly responsive and adaptive urban environment. At the heart of the Neom vision is an obvious shift in the way we conceptualize and interact with the built environment. Rather than static, inflexible structures imposed upon us by planners and architects, the cities of the future will be responsive ecosystems that evolve with their residents. AI and other advanced technologies will serve not as mere tools but as cocreators in the ongoing process of shaping and reshaping the cities in which we live. These systems will bring the idea of dynamically allocated resources from the realm of software programming to the physical world.

One only has to look at another Saudi project, called New Murabba, to extrapolate the extent to which spaces can be digitally customized. New Murabba is a futuristic downtown being planned just north of Riyadh that will accommodate up to 450,000 residents. At the heart of this new development will be one of the largest buildings ever built, the Mukaab. This 400m × 400m × 400m cubic structure—2.2 billion cubic feet!—will deliver immersive digital and holographic experiences by allowing its walls and ceiling to morph from one environment to another. Perhaps you see the ocean surrounding you one day, and on another you see thousands of hot-air balloons flying off into the clear sky. You are on Earth one morning and suddenly see the skies as they would appear from Pluto's surface on another. This massive structure will house a tower atop a spiral base, featuring high-end hotels, retail outlets, and cultural and tourist attractions.

Of course, the implications of Neom's cybernetic-city paradigm extend far beyond its technology and urban design. The integration of AI into the fabric of daily life will also have profound impacts on the social, economic, and governance structures that underpin society.

One way in which this can happen is with the integration of decentralized autonomous organizations (DAOs) into the governance fabric of the city. After decades of relatively mild changes in management theory, DAOs could represent a real shift. If they can work at scale, they would truly mark a step forward in organizational management, leveraging blockchain technology and smart contracts to create transparent, efficient, and democratic systems.

DAOs operate on the principle of merging human and machine decision-making into a cohesive framework. At their core, DAOs use "smart contracts"—self-executing computer programs that live on the immutable, unchangeable blockchain and automatically enforce terms of agreements when specific conditions are met. This integration ensures transparency and impartiality, significantly reducing risks associated with corruption and bias.

Consider, for example, a simple employment agreement managed by a DAO. In a traditional setting, an employment contract might specify that an employee will receive a bonus if they complete a certain number of projects by a specific date. But then there's the issue of whether the employer will keep their word, whether the goalposts will shift, and all manner of other doubts that can creep into situations like these. Implementing this with a smart contract on a blockchain transforms the process significantly.

First, the employer creates a smart contract that clearly outlines the bonus criteria. For instance, the contract might state: "If Employee X completes five projects by December 31, 2025, then release the bonus of $5,000." This contract is then deployed on the blockchain, where it becomes a permanent and unalterable record.

Next, the smart contract actively monitors the completion of projects. This monitoring can be achieved by integrating the smart contract with project-management tools that automatically log project completions on the blockchain. Each completed project is recorded in a transparent and tamperproof manner, ensuring that all parties have access to the same information.

As the specified deadline approaches, the smart contract autonomously checks whether the condition—completion of five projects—has been met. This process is carried out without any human intervention, relying solely on the data recorded on the blockchain.

If the condition is satisfied, the smart contract executes the terms of the agreement by automatically releasing the bonus payment to the employee's blockchain wallet. This ensures that the reward is delivered promptly and without the need for manual verification. Conversely, if the condition is not met, the smart contract takes no action, and the funds remain with the employer.

This automated process ensures that the terms of the contract are enforced impartially and transparently, eliminating the potential for disputes or bias. The use of blockchain technology provides a permanent and accessible record of the agreement and its fulfillment, further solidifying trust between the parties involved. The smart contract represents a sort of cybernetic extension of both your manager and the company for which you work.

By using smart contracts, DAOs can implement various types of agreements ranging from simple employment contracts to more complex arrangements involving multiple stakeholders and conditional actions. This approach not only enhances efficiency but also frees up human minds from fear and doubt. That alone can open up new possibilities for decentralized governance and collaboration.

But let's think beyond basic contracts. To me, one of the key advantages of DAOs is their ability to address issues inherent in traditional hierarchical structures, especially the issue of centralized power. In a DAO, decision-making power is distributed among all participants by promoting a flat organizational structure in which each member has a say in the direction and actions of the organization. This is typically done through blockchain-based voting mechanisms, where members use tokens—their piece of ownership in the DAO—to cast votes on proposals, ensuring transparency and consensus-driven outcomes. Importantly, the extent of a participant's ownership in the DAO is not just a function of money or the ability to purchase tokens. It can also be significant just based on how early that someone got involved in the project. In other words, people who care about the cause are more likely to have greater ownership. They can also earn governance tokens by carrying out acts of service for the DAO. And this "ownership-weighted democratization" of decision-making eliminates the need for a concentrated group of elites to hold power, fostering a more equitable environment where everyone can contribute to and influence outcomes.

The transparency inherent in DAOs is achieved through the public recording of all transactions and interactions on a blockchain. This immutable and auditable trail of activity enhances trust among participants and ensures greater accountability. Any attempts to manipulate the system or engage in malicious behavior can be quickly identified and addressed by the community.

The integration of DAOs into the governance structure of cybernetic cities like Neom offers a promising avenue for creating more equitable, efficient, and adaptive systems. For example, when it comes to making trade-offs among convenience, service, and environmental impact, why not use structures such as DAOs to determine where to build, how much to emit, and how to price the resulting product, whether energy or transportation? After all, the citizens of a city are the ones who have to live with such trade-offs, but they are almost never involved in the decision-making. Only after they feel the negative impacts, and only after they organize, campaign, and petition, might their voices be heard. Sometimes, it's too late. By harnessing the collective intelligence of both humans and machines, DAOs can revolutionize decision-making processes, resource allocation, and coordination of actions on a massive scale, enabling democratic governance, where all residents have a voice in shaping the community's future.

The use of smart contracts can streamline administrative processes, reduce bureaucratic inefficiencies, and ensure that public resources are managed effectively. And the transparency and accountability provided by blockchain technology can enhance trust between citizens and the government, creating a truly engaged and participatory society. To get a sense of how exactly Neom might use smart contracts, let's look at Estonia's e-Residency program, which leverages blockchain technology to allow global citizens to establish and manage businesses remotely. In this system, smart contracts automate key tasks like company registration, tax reporting, and compliance checks. The smart contract governs variables such as identity verification, business type, regulatory requirements, and tax thresholds. Once predefined conditions—such as completing the identity-verification process or meeting tax-filing deadlines—are met, the contract automatically executes the necessary actions, such as registering the business with government authorities or submitting tax filings. This automation reduces human error, eliminates the need for intermediaries, and ensures compliance with Estonian regulations in a transparent and secure manner. This concept can likely be taken much further, tying in utility-bill payments, bank accounts, ID verification for almost any social service, scheduling of medical visits with testing and insurance prerequisites, and more. A web of smart contracts can automate the most mundane and boring parts of life.

From a social perspective, Neom will likely produce significant changes in the way we live, work, and interact with one another. As the boundaries between human and machine agency blur, we'll ask important questions about the role of technology in shaping our social norms, values, and behaviors. We might look at these as small decisions—automatically checking into a hotel or automatically buying a train ticket if that happens to be the most convenient way to transit at a certain point in time—but this agency over making economic decisions being outsourced to algorithms will have effects much like ripples in a pond. They will begin to shape which infrastructure services, which companies and hotels are popular and which are not. These companies, in turn, will be analyzing their data and the data on competitors to determine why they are no longer popular. The sensors, DAOs, smart contracts, and other digital mechanisms in place in Neom will give them more data than ever before to analyze. I have a feeling that the only ones that ultimately survive will be those that respond fastest to changes in demand with a change in services and products. One can see where this naturally leads: to a place where the most competitive companies are the ones that analyze and respond in the fastest way possible, conceivably even with real-time offers, near-real-time changes, and customizations in products and repackaging of services.

Crucially, this focus on speed means that humans will likely stay out of the loop to the greatest extent possible. They may still play a role on boards of directors, providing high-level strategic guidance, but the adaptation on the ground may happen too fast for humans to remain too involved. There will be those who say this doesn't apply to specialized, artisan-like businesses. But even if so, those businesses are, by definition, niche. A small part of what matters. In fact, we've seen a preview of this focus on data-driven analysis and resulting adaptation on the internet. Today, a search engine like Google essentially controls the popularity of digital-media properties such as news or e-commerce websites. A small change in Google's algorithm can wreck a business or make it go viral. Today, manual analysis and automated tools are both used to constantly monitor rankings and reverse engineer Google's algorithm to try to figure out which micro-optimization in content results in a beneficial change in Google's ranking. Now imagine this type of analysis and adaptation using data from all sorts of sensors, text logs, video cameras, audio recordings, internet postings, blockchain transactions, and drone imagery.

Consumers might conduct microtransactions in the physical world based on recommendations made by their personal AI agents, and cities, companies, and countries might respond with maximally autonomous decision-making systems of their own, ultimately creating and consuming services made for humans. This is one small view into how "machine beliefs" manifest into the physical world as real products, services, neighborhoods, and even ways of life.

But we've also seen enough of the downsides of Big Tech to know that there are issues with increasing our reliance on AI in social contexts. The erosion of privacy and personal autonomy, the perpetuation of algorithmic biases and discrimination, and the potential for AI systems to influence our behavior in detrimental ways are all valid concerns. Yet the optimist in me wants to think about how we reframe these concerns not in light of a loss of autonomy but as an increase in our abilities as we morph from purely biological creatures to cybernetic beings. What if instead of outsourcing our personal AI agents to Big Tech, we can configure, adapt, and customize these systems on our own? What if these AI systems and their increasing agency in natural language and communication become a way to tear down the barrier of technological difficulty, allowing ordinary people to "program" these systems using nothing more than normal human speech and ordinary language?

There is reason to be optimistic on this account. The miniaturization of AI technologies means we are closer and closer to having useful edge AI: not Big Tech AI that needs billions of dollars in GPUs and terawatts of power to run, but miniaturized, embedded AI that lives and works locally on devices you own outright. Improvements in edge AI will enable us to carry our personal AI agents with us. The increasing sophistication of LLMs and generative AI in understanding and communication means that these AIs will be able to learn about us and respond in highly tailored ways. Because these devices will be carried by us and because they will store data locally on equipment we own outright, we will trust them more than we do cloud-connected devices that siphon our information off to Big Tech companies. Consequently, we will allow these trusted, owned devices to know more about us than we might be comfortable doing today. And by knowing more about us, perhaps even mirroring our senses—seeing what we do, hearing what we do—these devices will be immensely more useful. They won't answer general questions targeted toward a single user demographic, such as men in the southern United States. They will answer the question just for me, based on what I've seen visually, what I've heard, and what I've read. And I will no longer fear being bent to some tech overlord's will. The device is mine, the data are mine, and I own it all outright. Any algorithm that runs on my device does so with my permission, and I can turn it all off in whole or in part whenever I choose.

With such trusted, edge AI devices, the personalized, data-driven services and experiences we imagine existing within Neom could in fact foster greater social connection and cohesion, enabling individuals to access insights and resources that enhance their well-being and sense of community. It is possible to take responsibility for our technology as individuals and, rather than allow a central authority to control or shape our decisions, become nodes in a super-empowered knowledge network, seamlessly sharing knowledge, experiences, and resources across our diverse communities. How might we create new opportunities for collaboration and mutual understanding if we can pull this off?

In economic terms as well, the rise of cognitive cities will have major implications for the global landscape of innovation and entrepreneurship. As hubs for cutting-edge technologies and industries, these cities will likely attract significant investment and talent from around the world, potentially fostering the emergence of new economic clusters and innovation ecosystems. Much as London, New York City, and Silicon Valley became magnets that attracted the world's best and brightest, countries that build successful cognitive cities will become a magnet for talent.

AI for urban infrastructure and services will also drive significant gains in productivity and efficiency, unlocking new sources of economic value and creating new opportunities for growth and development. Ensuring that the benefits of this technological revolution are distributed equitably and do not exacerbate existing inequalities will not just pose a technological challenge, in terms of making the tech easily accessible to all, but will also be a question of political understanding and leadership. Countries with politicians who understand technology will be better suited to this shift and will adapt to this new era more successfully.

As I write this, I can see the video of Mark Zuckerberg's testimony to the US Congress playing in my mind. The eighty-four-year-old Utah senior senator, Orrin Hatch, clearly clueless about how social networks work, asks Zuckerberg, "So, how do you sustain a business model in which users don't pay for your service?"

"Senator, we run ads," replies Zuckerberg, with a smirk.

A Senate full of people with Senator Hatch's seemingly nonexistent understanding of technology will be a liability in the era of cognitive, cybernetic cities.

There will be challenges to address. The integration of AI into urban governance and decision-making will raise complex ethical and political questions. How will the use of AI in urban planning and policymaking be regulated and overseen? How will citizens' rights and interests be protected in a world where algorithms play a significant role in shaping daily life? And perhaps most fundamentally, how do we ensure that the values and principles underpinning these powerful technologies remain firmly rooted in the ideals of human dignity, freedom, and self-determination?

When decisions concerning resource allocation, public-service delivery, law enforcement, transparency, accountability, and democratic oversight become cybernetic decisions, how will we ensure that they deliver a better society for all citizens? Do such decisions simply become an optimization problem where they are made, then their outcomes measured constantly and inputs adapted until we maximize some measure of social value? And what will that value be? Will democracies evolve to the point where we vote on selecting what we want our cybernetic systems and cities to optimize?

These are not just hypothetical concerns. The decisions and frameworks established in these pioneering urban environments will set important precedents and benchmarks for how societies around the world navigate the integration of AI and other transformative technologies into the fabric of civic life. I can see the idea of democracy itself changing and evolving in these cities of the future.

Today's macroscopic democracy provides what increasingly seems a mere illusion of choice. Particularly in the United States, we are forced to choose between candidates who most people don't much care for. As of October 2024, both Vice President Harris and former president Trump have net-negative ratings, according to the Pew Research Center. This means that more people dislike them than like them. Charitably, one could argue they are an average of averages. On darker days I imagine that our election has devolved to a selection: a selection by the party apparatus and layers and layers of bureaucracy that might be more interested in protecting itself than in ensuring the will of the people. No matter how you see it, I hope you will agree that the prospect of using technology to evolve our democracy into one where people are invited to increase their participation with greater convenience is a good thing. This would be a democratic system that uses technology to seek the counsel of its citizens far more frequently on more immediate and even smaller-scale matters that actually affect our day-to-day lives. And for those who might ask whether people would even care about each micro-decision, perhaps they will. And if they don't, perhaps they will use voting smart contracts to constrain their overall choices based on high-level guidance and then examine the immutable, blockchain-hosted manifesto of each candidate to identify one who comes closest to fulfilling their criteria as a whole. We don't have to let the machines take over. We can also use machines to take back the choice we humans have lost.

And these social innovations will extend beyond community governance to every scale of human life. For example, the environment. Neom's emphasis on sustainability, ecological harmony, and the seamless integration of natural and built environments is inspiring. The city's designers envision a metropolis that exists in symbiosis with its surrounding ecosystems, minimizing its environmental impact while creating opportunities for residents to connect with and appreciate the natural world.

From expansive urban greenways and re-wilded natural spaces woven into the city's fabric to cutting-edge technologies that enable the regeneration and restoration of local habitats, Neom aims to build a bridge between urbanization and the environment. The goal is not just to mitigate the negative impacts of human activity but also to actively enhance and enrich the natural world through our collective ingenuity and stewardship.

But beyond its commitment to environmental sustainability, Neom's vision for human flourishing extends to the realms of health, education, and personal growth. By leveraging the power of AI and advanced technologies, the city aims to create a personalized, data-driven ecosystem of services and experiences tailored to the unique needs and aspirations of each individual.

I've met with several of the brilliant people working at Neom. I've spoken to them about how their work is unfolding. And about their collective vision for the place. I've sensed the optimism in their voices and in what they tell me. But as I research Neom on the internet, I also see lots of skepticism and, strangely, some jealousy. Many in the United States and Europe in particular lampoon the idea or prophesy that it will never come to pass. I ask myself, "Why this skepticism?" Is it genuinely rooted in an understanding of large-scale project management, of artificial intelligence or urban planning? Perhaps 1 percent of it might be. But the vast majority is negativity simply because of where Neom happens to be. Simply because of who is building Neom. "Why can't we have this?" is what I imagine is the unstated sentiment behind much of the caustic commentary.

We have to move beyond this us-versus-them view of the world. Before we land on Mars, before we build colonies on the moon, we have to at least begin to imagine ourselves as the human race, united in the pursuit of worthwhile challenges. If we want it to be, Neom can be not just a city but a global symbol as well—a testament to the potential of human ingenuity and to our enduring capacity to reshape the world around us in ways that transcend the imagination of our forefathers. If we want it to be, it can become a beacon of hope and possibility.

CYBERNETIC HOUSES OF WISDOM

The House of Wisdom, or *Bayt al-Hikmah*, was established in Baghdad during the reign of Caliph Harun al-Rashid in the late eighth century CE. This intellectual center became a beacon of knowledge during the Islamic Golden Age, housing scholars from various cultures and religions who engaged in extensive research, translation, and education. It served as a repository for the collective wisdom of the ancient world, incorporating texts from Greek, Persian, Indian, and other traditions, which were translated into Arabic and further developed by resident scholars.

Under the reign of Caliph al-Ma'mun, the House of Wisdom flourished even more. Al-Ma'mun, an avid supporter of science and philosophy, expanded the institution by inviting renowned scholars to Baghdad and funding their research. The House of Wisdom became the epicenter of scientific and philosophical inquiry, leading to significant advancements in fields such as astronomy, mathematics, medicine, and chemistry. Scholars like Al-Kindi and Al-Khwarizmi made groundbreaking contributions, including the development of algebra and the introduction of the Arabic numeral system, which laid the foundation for modern mathematics.

The architectural design of the House of Wisdom reflected its grandeur and purpose. It featured a large building with numerous halls and rooms dedicated to different fields of study and activities such as translating, copying, and binding books. The facility included an observatory that facilitated astronomical research, enhancing the scholarly output of the institution. The House of Wisdom was not only a library but also a vibrant intellectual hub that significantly influenced the cultural and scientific landscape of its time, originating the spirit of innovation that was later echoed in the Renaissance.

The impact of the House of Wisdom extended beyond Baghdad. It inspired the establishment of similar institutions across the Islamic world, such as Dar al-Ilm in Cairo, founded in 1004 CE by Caliph al-Hakim. These centers of learning became pivotal in preserving and expanding upon ancient knowledge, fostering an environment where scholars of different backgrounds could collaborate. This widespread network of intellectual hubs significantly contributed to the scientific and cultural advancements of the Islamic Golden Age, making the era a cornerstone of global intellectual history. Just as the House of Wisdom had an impact beyond Baghdad, might Neom have an impact far beyond the shores of the Red Sea?

The methodologies developed at the House of Wisdom, such as systematic experimentation and empirical observation, laid the groundwork for the modern scientific method. Scholars like Ibn al-Haytham, known for his work in optics, exemplified this approach. His *Book of Optics* fundamentally changed the understanding of vision and light, influencing both Islamic and later European science. This spirit of inquiry and rigor in research was crucial in transforming theoretical knowledge into practical applications, which benefited various fields, including medicine, engineering, and astronomy.

The legacy of the House of Wisdom can even be seen today in contemporary institutions that strive to integrate diverse fields of knowledge and foster interdisciplinary research. Modern equivalents include research centers like the Massachusetts Institute of Technology (MIT) and the King Abdullah University of Science and Technology (KAUST) in Saudi Arabia. These institutions are trying to create environments where the convergence of different disciplines can lead to innovative solutions to global challenges. They embody the enduring influence of the House of Wisdom by continuing to promote the values of collaboration, open inquiry, and pursuit of knowledge for the betterment of society.

Ultimately, a global cybernetic-city network's capacity for intensive cross-pollination and combinatorial novelty could echo and amplify the ideological-fertilization effects once produced by the seminal melting pots of ancient Greece, the House of Wisdom, or the European Renaissance's catalyst cities. This superlinear innovation ecosystem, turbocharged by epicenters like Neom, could potentially produce a flourishing redefinition of civilization itself. Neom, as not just a center of learning but as an entire city dedicated to collaboration and to inventing the future, might leave behind an even greater legacy than the great historical centers of learning.

Our ancestors achieved a grandiose elevation of civilization when they built the first city-states and then the global urban centers that served as centers for intellectual exchange and fusion. With all the technology at our disposal, is it really such a miraculous idea?

Can we not attempt the same?

The choice is ours.

CHAPTER 3

COMPANIES AS CYBERNETIC ORGANISMS

The year is 2035, and the best performing stock of the past nine months is NTKX, a company no one had heard of a year ago. From founding to going public, it was an absolute rocket ship, the first of its kind: a cybernetic corporation.

NeuralTech (NTKX) has become a global leader in genomics, looking not only at the genetic material of organisms but also at how that information is applied. The company has made advances in the medical field, studying drug responses and disease. New discoveries have been made in microevolution, which has helped identify how environmental factors are influencing populations. And its work in synthetic biology has transformed everything from bioprinted organs to the food we ingest. No matter where we look in the field of genomics, NeuralTech is right there at the forefront.

Unlike traditional companies of the early twenty-first century, NeuralTech operates on a fundamentally different paradigm, one that seamlessly integrates human intelligence with the boundless capabilities of AGI (artificial general intelligence) systems and autonomous robots.

At NeuralTech, the traditional hierarchical structure of companies stands obsolete. The central intelligence core (CIC), the advanced decision-making and planning AI system at the heart of the company, has the ability to directly manage and interact with every single employee, regardless of their role or location. This is in stark contrast to the limitations faced by human managers who, in previous eras, were always constrained by their cognitive limits, able to handle 5 people in the closest circle of trust and no more than 150 with whom they could maintain stable social relationships. The limitations of human managers resulted in hierarchies, and hierarchies brought their own problems: inefficiency, infighting, disagreement, and disaffection.

The CIC's ability to transcend such limitations has had profound implications for the organizational structure of NeuralTech. Instead of pyramid-shaped hierarchies with layers of middle management and executives, the company operates on a flat, decentralized model. Every employee, human or machine, has a direct line of communication with the CIC, which assigns tasks, provides feedback, and offers support in real time, all the time.

This flattening of the organizational hierarchy unlocks unprecedented levels of efficiency and agility for NeuralTech. Without the need for intermediaries or bureaucratic red tape, information flows freely and decisively across the organization. The CIC optimizes resource allocation, identifies and resolves bottlenecks, and adapts to changing circumstances with a speed and precision that would be impossible for a human-led company. It turns out that in a company made up of humans and machines, the most efficient thing to do is to make the machine the CEO!

The CIC's ability to manage a vast network of employees also enables NeuralTech to scale its operations in ways that traditional companies could only dream of. The company takes on projects of immense complexity and scope, leveraging the collective intelligence of its human and machine workforce to solve problems that were previously insurmountable. The CIC easily identifies which employees are best suited to work on which projects, given skills, bandwidth, and work–life considerations. This paves the path for an overall higher level of satisfaction from employees across the board. In addition, the CIC keeps tabs on when motivation and productivity may be slipping. When identified, issues are addressed at an individual level. What motivates one individual may not do the same for another. And when personal information is shared with the CIC, the information is kept confidential. The CIC doesn't let slip who's having trouble at home or who's scheduled for what medical procedures. And when employees complain of job monotony or fatigue, the CIC looks for new opportunities to respark interest.

This organizational transformation has had a remarkable impact on the company's bottom line. Its productivity and innovation have soared, allowing it to outcompete traditional firms in almost every industry it enters. NeuralTech is entering new industries at a breathtaking pace! The CIC's data-driven decision-making and real-time adaptation have turned out to be an asset in even the most dynamic and unpredictable markets.

A group of short sellers, hoping to make a gain from the stock falling, launches a vicious attack on NeuralTech just three months into its IPO, claiming that the company is running afoul of "the human interest" and that its decisions are unethical and not in the interest of society. But the CIC was a step ahead . . .

NeuralTech had already implemented a range of safeguards and ethical guidelines to ensure that the CIC's power is used responsibly and transparently. The CIC had predicted the risk of a short-seller attack and had established an independent ethics board, composed of both human and machine representatives, to oversee the CIC's decision-making and ensure that it aligns with human values and interests. All this information had been kept hidden because the CIC knew that releasing it beforehand would only make it irrelevant when the inevitable short-seller attack occurred. Instead, the information is released on the day after the short-seller attack, along with an established history of performance. This effectively neutralizes the negative press, making the short sellers seem uninformed. The attack is thwarted. But the CIC doesn't become complacent. It prepares for the next such potential attack.

The rapid ascent of cybernetic corporations like NeuralTech has sent shockwaves through the business world. Veteran CEOs, worried about their lucrative positions and massive bonuses, are in open revolt. Human-run, traditional companies have struggled to adapt to this new flat-organization paradigm, finding themselves outmatched by the speed, efficiency, and scalability of their cybernetic rivals. Some have attempted to retrofit their organizations with AI and automation, but without the same level of integration and sophistication of a "CEO" like CIC, they have found it difficult to compete. As NeuralTech continues to expand into new markets, these traditional companies face a true threat to their existence.

NeuralTech's meteoric rise has made it abundantly clear that the future of business belongs to those who can successfully harness the power of artificial intelligence and human–machine collaboration, even if the humans must report to the machines. The company's success has inspired a new generation of entrepreneurs and innovators who are racing to build the next generation of cybernetic corporations with AI as their CEO.

Back to reality . . . but the core hypothesis at the heart of the story, that a truly cybernetic corporation will be flat, and will be able to overcome the cognitive limitations of humans, deserves serious consideration. Under Jensen Huang's leadership, Nvidia has become the world's largest corporation, and it provides an interesting case study. In multiple interviews, Huang has emphasized the importance of Nvidia's flat organizational structure. In a 2021 interview, he stated that "I have about twenty direct reports. I believe in a very flat organization." He elaborated on this philosophy in another interview: "The company is as simple as possible so that information can travel as quickly as possible." Not every CEO has the capacity to deal with twenty direct reports. But a CIC might be able to manage many times the number of people Jensen Huang does today. So is NeuralTech or its equivalent about to emerge, whether we like it or not?

It is certainly not out of the realm of possibility. But first, what exactly is a cybernetic company? A cybernetic company is an organization that functions as a self-regulating system, seamlessly integrating human intelligence with artificial intelligence and advanced automation to operate at unprecedented scale and efficiency. In such a company, decision-making is distributed not through traditional hierarchies but via an advanced AI core that manages and directs the company's operations; the company is extremely "flat."

This AI core acts as a central nervous system, continuously processing vast amounts of data and making real-time decisions on everything from resource allocation to employee well-being. All the information necessary for such decisions to be made is gathered via a number of sensors, including visual, infrared, text, audio, and any others that might be developed. The cybernetic company transcends the limitations of human cognition by leveraging the speed, precision, and tireless capacity of AI systems. Humans and machines collaborate under the guidance of this AI, which optimizes workflows, solves complex problems, and ensures that ethical considerations are built into its operations.

In essence, a cybernetic company is a decentralized, intelligent entity where the boundaries between human and machine blur, allowing it to operate with agility, scalability, and adaptability far beyond the capabilities of traditional, hierarchical corporate structures in which human decision-making and structural inefficiency cause significant delays in almost all tasks and actions.

Fulfilling such a definition in absolute terms will take time, but in our own reality, companies are already on track to morph into such cybernetic corporations. And with strong incentives: The transformative integration of AI into organizational ecosystems will not just be an evolution in technology in the sense of a human using a passive tool, but a revolution in the collaborative dynamics between humans and machines. This time, the "tools" will think as well, not just the humans. Organizations at the forefront will recognize AI as a dynamic partner in their operational and strategic endeavors—a perspective of symbiosis where AI and human workers learn from and teach each other, propelling companies toward previously unattainable levels of success.

For organizations to thrive in this era of rapid technological advancement, fostering a culture of collaboration between humans and machines will be crucial. By viewing machines as partners, organizations can gain access to a flywheel of innovation, leading to improved effectiveness and well-being. Leaders should design workflows that capitalize on the complementary strengths of humans and machines, thereby creating more intelligent, adaptive, and resilient organizations.

Deloitte Insights sheds light on how AI revolutionizes infrastructure management across industries. Through remote sensing and artificial intelligence, companies improve operational efficiency and accuracy while minimizing the need for human intervention in monotonous tasks. This allows human employees to focus on more complex and strategic decision-making. McKinsey & Company emphasizes that integrating AI into teams can lead to significantly improved outcomes, with health care being a prime example. AI's data-analysis prowess enhances diagnostic processes, allowing doctors to focus on patient care, empathy, and treatment planning.

Integrating AI into organizational operations involves not just technical implementation but also a reevaluation of work environments, rituals, norms, and collaboration. It calls for creating spaces where humans and machines can work together seamlessly, leveraging each other's strengths to solve complex problems and innovate.

PROTO-CYBERNETIC COMPANIES

In the global e-commerce market, Amazon is not merely a titan in revenue terms; it is also the driver of a nascent cybernetic revolution that blurs the once clear demarcations between human intellect and machine automation. Amazon's founder, Jeff Bezos, with his relentless drive toward operational efficiency and obsession with consumer satisfaction, has spearheaded major efforts to bring about a synergy between human endeavor and artificial intelligence, positioning Amazon at the forefront of a cybernetic paradigm shift. Bezos has said that he sees the human brain as "an incredible pattern-matching machine." And he also speaks about innovation in a combinatorial context: "If you double the number of experiments you do per year, you're going to double your inventiveness." This is the type of argument that underscores why machines can out-innovate. It is because they can out-compute, out-experiment, and out-invent humans with sheer speed and parallelism.

The more you listen to Bezos, the more you realize that he sees Amazon as a cybernetic construct made up of both humans and machines that must be optimized, not in its parts, but in its sum. He is also not afraid of criticism, which is easy to receive when you move the slider between the number of humans and the number of machines at a company. Moving that slider has consequences in the real world. It means fewer or more jobs. Higher-paid or lower-paid jobs. It matters to human beings. But Bezos also sees stasis as a threat: "I believe you have to be willing to be misunderstood if you're going to innovate. . . . What's dangerous is not to evolve, not to invent, not to improve the customer experience."

Amazon's sprawling facilities, where 1.5 million human employees globally join forces with a formidable assembly of 750,000 robots, provide a living example of this innovation. These robots do everything from picking products up from shelves to lifting and carrying bulky items. One robot, named Titan, can lift up to 2,500 pounds. It is estimated that 75 percent of customer orders are delivered with the assistance of robots. Although it may seem that these robots are entirely replacing humans, Amazon has created an estimated 700 new categories of jobs since robots were introduced to the workforce in 2012. It has even established the Mechatronics and Robotics Apprenticeship Program, which helps employees gain new, more marketable skills. According to Amazon, those who graduated from the program have had an estimated 40 percent increase in pay.

As a company, Amazon has not only used but also built many of these robots. What the company hasn't built, it has bought. To move its automation efforts faster, it acquired robotics businesses Kiva Systems and CANVAS Technology, signaling a departure from conventional labor paradigms toward a hyperefficient, robotics-infused future. Kiva, founded in 2003 and now renamed Amazon Robotics, was purchased in March 2012 for $775 million.

At the time, it was Amazon's second-largest acquisition. Amazon Robotics manufactures mobile-robotic fulfillment systems that in the past have serviced companies such as The Gap, Crate & Barrel, Walgreens, and Office Depot. CANVAS, which was founded in 2015 and made a splash in 2017 with the reveal of the world's first completely autonomous self-driving warehouse cart, was acquired by Amazon in April 2019.

Within Amazon's nurseries of innovation lie the Sequoia system and the bipedal humanoid robot known as Digit. Sequoia will drive advancements in inventory management and order processing, allowing inventory to be identified and stored up to 75 percent faster than before. In addition, the time it takes to process an order will be reduced by up to 25 percent with the help of this new system.

Digit resulted from a collaboration with the fast-growing start-up Agility Robotics and ventures into previously uncharted domains of logistical operations with its humanoid form and capacity for autonomous interaction with its environment. The first two robots rolled—or stepped—off the Agility Robotics assembly line in 2020, with Ford Motor Company as the first customer. The robots got right to work, helping improve warehousing and delivery for customers. With its founder-inspired mindset of continual improvement, Amazon soon partnered with the company. These technological leaps forward are just two examples of Amazon's relentless pursuit of the outer limits of current technological capabilities, all in the name of operational efficiency and even faster customer service.

So how is Amazon faring as it enters its cybernetic epoch?

The initiation of real-world trials for Digit within Amazon's BFI1 fulfillment center in Sumner, Washington, marked a significant milestone not just for Amazon but also for Digit's designer, Agility Robotics. Charged with tote consolidation, Digit has shown great promise in taking on a task that was heavily reliant on human labor. Its present phase of testing, as explained by Emily Vetterick, director of engineering for Amazon Robotics, forms a crucial part of a meticulous developmental cycle, emphasizing the collection of employee feedback so that Digit can seamlessly integrate with other robots and humans working at Amazon.

The humanlike appearance and capabilities of robots such as Digit introduce a new dynamic within the workplace, necessitating careful consideration of employee perceptions and the broader implications of anthropomorphic designs, the most obvious being "If they look like us, and outperform us, what remains of our value?" If you ask Amazon, they'll tell you that the integration of robots like Digit into their workforce underscores their commitment not to supplant human labor but to elevate it, transitioning from tasks deemed "dull, dirty, and dangerous" toward roles that are more intellectually fulfilling and less physically taxing. What's not to like about that? But this progression toward a symbiotic coexistence between human workers and robots is not without its challenges.

Beyond humanoids and robotic warehouses, Amazon also builds products that result from a delicate human–machine symbiosis. Amazon's Mechanical Turk platform allows customers to hire humans and machines all over the world to conduct simple tasks like data labeling, editing, spell-checking, reformatting, and collecting. Here, the veil between human and machine intelligence becomes translucent, presenting services that, while appearing automated, are powered by the cognitive efforts of millions globally. Amazon's "Turkers" engage in tasks that breathe life into AI and machine-learning models, creating future training data and showcasing the profound intertwinement of human and artificial intelligence. Will humans become useless once they've provided AI with the requisite training data? That remains to be seen.

Amazon has also implemented an algorithmic system to monitor and manage the productivity of its warehouse employees, a process that has led to automated firings without any human intervention. The system tracks the performance metrics of individual workers, measuring the speed and efficiency with which they complete tasks.

One of the primary areas where this system is applied is within the Amazon Flex program, which uses drivers to deliver packages. Flex drivers have reported being terminated by automated emails that cite violations of terms of service or inadequate performance. Often, these terminations come with little explanation or recourse. For example, according to Bloomberg, drivers have received conflicting communications regarding their performance status, only to be abruptly deactivated with automated responses when they seek clarification. This has led to severe financial and personal hardships for many workers, who find themselves without a job and with no clear understanding of the reasons behind their termination. Although the goal may be to motivate employees and keep them aware of their performance, the implementation seems to have its flaws.

Stephen Normandin, a sixty-three-year-old army veteran, was fired via an automated email from Amazon after working for four years delivering packages for the company in Phoenix, Arizona. Critics, such as Stacy Mitchell, codirector of the Institute for Local Self-Reliance, argue that this system dehumanizes the workforce, treating employees as mere cogs in a giant machine. The reliance on algorithms to make significant employment decisions highlights a disconnect between the realities of workers' lives and the cold efficiency required by managers and implemented through machine logic. This approach can overlook the complexities and nuances of human performance, such as unexpected emergencies or challenging personal circumstances, leading to high turnover rate, violations of workers' rights, and unnecessary impacts on job security.

In this new cybernetic reality, the relative importance and value of human and machine components are under question. What is fair compensation? What does job security look like? And what are the rights of an often-invisible workforce? The broader implications of Amazon's use of AI for employee management extend beyond individual cases.

Amazon's "Hands off the Wheel" program is already automating various corporate tasks, including forecasting, pricing, and purchasing. An internal software tool called AC3 is focused on customer service and provides a Q&A dialogue for service agents to follow. Data captured with AC3 can be used to train AI models that can potentially replace human agents. Amazon has also been looking to extend the use of AI and machine vision to monitor performance. In fact, it has already run into trouble because of the extent of its monitoring. In December 2023 the French data-protection authority fined Amazon €32 (US $33.5) million for "excessively intrusive" surveillance of warehouse workers, including monitoring work interruptions and breaks too closely.

All this begs many questions. If an employee falls short of productivity standards set by the algorithm, how should demands for efficiency stack up against significant concerns about fairness and transparency? After all, how is a machine supposed to take personal considerations into account? Even the most motivated employees may have subpar days, or weeks. How does empathy get inserted into the equation?

Today, Amazon is achieving higher performance by melding together human cognitive abilities with the computational power of machines and the physical capabilities of robots. How the ratios among humans, robots, and other AI-enabled systems evolve will tell us a lot about the future. By watching Amazon, we are not just keeping up with a cutting-edge, tech-enabled company with a brutal focus on profitability and efficiency. We are also keeping our eyes on whether we can expect an equitable and sustainable future that harnesses the strengths of both human and machine intelligence.

Accenture predicts that companies that "put people at the center, and commit to develop responsible AI systems," could improve profitability by an average of 38 percent by 2035. The expansion of Amazon's robotic workforce is an early test of how significant the socioeconomic aftershocks of this cybernetic evolution will be. Even as such questions are asked, Amazon propels forward with its automation endeavors and, in doing so, widens the gap between itself and its US-based competitors. Because robots are still an expensive investment, it is unlikely that smaller entities will be able to invest in the technology necessary to achieve the hyperefficient human–machine symbiosis that Amazon is aiming for. Just as they somehow survived the e-commerce revolution, it is possible that niche sellers will remain viable amid disruptive technological advancements, but they will have no chance of broad competition with cybernetic giants like Amazon and will be unable to match Amazon's customer-response times and prices.

AI AND SCALING LAWS

What would happen if companies could rely on AGI (artificial general intelligence) systems to make decisions, and on a mix of humans and robots to execute those decisions? For starters, the process by which companies evolve would likely undergo significant transformations. Fundamental to these changes are what are known as laws of scale.

Physicist and Santa Fe Institute researcher Geoffrey West explains and analyzes these laws in his excellent book, *Scale*. West's work has its origins in the attempts in the 1930s to describe how various characteristics of animals related to one another; the classic work of the early twentieth century sought to link, for example, an animal's basal metabolic rate to its body mass. That work led to the formulation of a rule known as Kleiber's Law, which says that across species, metabolic rate scales as body mass to the power of 3/4. Thus, the law predicts that a mouse has a faster metabolic rate per gram of mass than an elephant does.

West and others have explored how these mathematical relationships can predict various attributes of biological systems based on size, such as metabolism, lifespan, and growth rates. West found similar predictable scaling regularities in various urban metrics. For example, he found that productivity and innovation scale at faster than linear growth when plotted as a function of city population. Infrastructure needs, on the other hand, scale sublinearly as a function of population. These relationships give us predictable, usable information about the growth patterns of human-made systems, including industries, corporations, and municipalities. For instance, if we want a city to be highly innovative, we are better off making it large.

It is remarkable that similar scaling laws appear in both biological systems and human-made structures, suggesting fundamental principles governing complex systems across vastly different domains. West and his colleagues have extensively studied these scaling laws, proposing that the constraints limiting the size of animals may also apply to cities and other human-made systems.

In biological organisms, the circulatory system faces diminishing returns as it scales. Blood vessels branch repeatedly, eventually reaching a point where maintaining sustainable pressure throughout the system becomes challenging without risking damage to capillaries. West's research shows that metabolic rates in organisms scale as a 3/4 power of mass, a relationship that holds across 27 orders of magnitude, from molecular to organismal levels.

Similarly, cities encounter limitations in their distribution networks—roads, electrical grids, and communication lines—which, like blood vessels, distribute vital resources and information. These networks, though efficient at smaller scales, become increasingly strained as cities grow. However, cities differ from biological organisms in a crucial way: they exhibit both sublinear scaling in infrastructure and superlinear scaling in socioeconomic outputs.

West and his colleagues have found that as cities grow in size, they become more efficient in terms of infrastructure, requiring less energy and resources per capita. Simultaneously, they become more productive and innovative. When a city doubles in size, key metrics such as economic productivity and patents increase by approximately 15 percent more than if they scaled linearly with population. Don't take this to mean that cities can scale indefinitely or that it would be good if they did. Greater efficiency in this context means only that for every subsequent unit of growth in city size, slightly less infrastructure is required. However, it doesn't mean that for a large enough city, no incremental infrastructure would be required. Nor does it mean that the size of infrastructure needed by a city much larger than any we know today would be sustainable. There are other challenges that arise as cities scale.

The phenomenon of greater productivity is quite likely rooted in the increased interactions and social connectivity that larger cities facilitate, leading to the emergence of new ideas, technologies, and economic opportunities. The density of cities intensifies their capacities per capita, fostering innovation and economic growth that offset some of the physical inefficiencies in infrastructure.

However, the density of cities also presents significant challenges of other varieties. West suggests that the overwhelming stimuli in dense urban environments can lead to a kind of cognitive overload, potentially affecting human behavior and ethics. This idea is rooted in earlier research, particularly a famous experiment conducted by psychologist Philip Zimbardo in 1969. Zimbardo abandoned two cars in different locations: one in the Bronx, New York, and another in Palo Alto, California. The car in the Bronx was vandalized and stripped of parts within hours, but the one in Palo Alto remained untouched for over a week. This stark contrast demonstrated how environmental cues and social conditions can significantly affect behavior, forming the basis for the "broken windows" theory of crime.

The broken windows theory, introduced by social scientists James Q. Wilson and George L. Kelling in 1982, posits that visible signs of disorder and neglect in a neighborhood can lead to more serious crime. The theory suggests that minor infractions, if left unchecked, signal that no one cares about the area, potentially encouraging more serious criminal behavior. Maintaining and monitoring urban environments to prevent small crimes helps create an atmosphere of order and lawfulness, thereby preventing more serious crimes.

West's interpretation of this phenomenon in the context of urban scaling suggests that as cities grow denser, the increased stimuli and anonymity might contribute to a breakdown of individual ethics similar to the accelerated vandalism observed in Zimbardo's Bronx experiment. This dynamic illustrates the complex interplay among urban density, social behavior, and the challenges of maintaining order in rapidly growing cities.

The relationship between the broken windows theory and city size is not direct but rather a consequence of how cities scale and the emergent properties of larger urban environments. As cities grow larger, the sense of anonymity among residents increases, potentially making it easier for minor infractions to occur without social consequences. The higher population density in larger cities leads to more frequent interactions and potential conflicts, as well as more opportunities for minor infractions to occur and be observed.

Larger cities have more extensive infrastructure to maintain, and although they benefit from economies of scale in some aspects, the sheer volume of urban space to monitor and maintain can make it challenging to quickly address signs of disorder. These cities often have greater socioeconomic diversity and, frequently, more pronounced inequality, which can create areas of concentrated disadvantage where signs of disorder may be more prevalent and harder to address.

West's research shows that many social phenomena, including crime, scale superlinearly with city size. This means that as cities grow, these issues tend to increase at a faster rate than population growth, potentially making the broken windows effect more pronounced. Probabilistically, a single ignored instance of crime or a delayed intermediation in the event of a "broken window" is more likely the larger a city gets. And if each broken window invites more broken windows, this effect is likely to build faster in a larger city than a smaller one. Additionally, larger cities have more complex governance structures, which can make it challenging to implement and maintain consistent policies across all neighborhoods.

Interestingly, whereas cities exhibit superlinear scaling in socioeconomic outputs, companies tend to experience sublinear scaling in both infrastructure and outputs. This contrast highlights the unique scaling properties of cities, which thrive on density and interaction, but companies struggle to maintain flexibility as they expand.

Evidence for this can be found in the Compustat data analyzed by West that covered 28,853 companies traded on US markets between 1950 and 2009. The data reveal a power law relationship between company size (measured by number of employees) and net income, with a scaling exponent of 0.79. This sublinear scaling means that as a company increases its number of employees by a factor of 10, its net income increases only by a factor of approximately 6, indicating diminishing returns to scale.

West's analysis shows that this sublinear scaling applies not just to net income but also to sales and gross profit. This suggests that as companies grow, they face challenges in maintaining efficiency and profitability across all aspects of their operations.

West proposes that this fundamental difference in scaling behavior explains why cities experience open-ended growth but the growth of companies eventually stalls. The sublinear scaling of company outputs suggests that they will eventually stop growing and ultimately die as they run out of profit-generating growth opportunities.

This difference in scaling behavior between cities and companies highlights the unique properties of urban environments, which seem to foster innovation and productivity in ways that large corporations struggle to replicate. It suggests that the dense, interactive nature of cities creates an environment more conducive to sustained growth and innovation than the more structured, hierarchical organization of large companies.

The sublinear scaling of infrastructure in cities can be understood through the lens of cybernetics, reflecting the optimization of resource allocation and efficient coordination between human and machine components. The integration of sensors, artificial intelligence, and data analytics in urban environments enables real-time monitoring and adaptation of infrastructure, potentially leading to smarter, more resilient urban systems. These technological advancements may help mitigate some of the psychological challenges posed by urban density, although they also introduce new complexities.

West's research provides a framework for understanding and potentially predicting the behavior of complex systems as they scale, offering insights that could be valuable in designing more efficient and sustainable cities and organizations. It also underscores the need for innovative urban planning and social strategies to address the psychological and ethical challenges that arise from increased urban density and complexity. As cities continue to grow and evolve, balancing the benefits of urban scaling with the need to maintain social order and individual well-being will be crucial for creating livable, sustainable urban environments. Technology for automated maintenance, infrastructure optimization, safety, and security management can all likely play a role.

From a cybernetic perspective, the failure of companies to scale effectively can be understood as a breakdown in feedback mechanisms and communication channels. Let's remind ourselves of Jensen Huang's view of this phenomenon, which we covered at the beginning of the chapter: "The company [Nvidia] is as simple as possible so that information can travel as quickly as possible." What Huang is saying is that as companies grow, they become more hierarchical and siloed, with information flowing less freely between different departments and levels of the organization. This leads to a disconnect between decision-makers and frontline employees, as well as between the company and its customers, suppliers, and partners. The result is a loss of agility and responsiveness, which can ultimately lead to stagnation and decline. The antidote to this is an assault on hierarchy.

If companies are to survive, they must grow. If they don't grow, they offer nothing of value to investors. And absent investor interest and availability of capital, the capital flows to their competitors, and they die. To continue to live, companies must keep scaling. Today, the largest companies hover around the three-trillion-dollar mark in terms of valuation. And as we've discussed, one of these giants is Nvidia, whose CEO is already dealing with twenty direct reports in an attempt to build as flat an organization as possible. Can he scale this to thirty-five or fifty direct reports? Unlikely. So if flatness enables communication, and communication is the key to marshaling resources effectively, which in turn is essential to growth, then flatness of structure should be an express objective. We seem to be at the limits of human leaders enabling flatness, so the logical conclusion is that in order to continue to grow, companies may have no option but to embrace a more cybernetic approach to organization and management, one that emphasizes the importance of flat hierarchies enabled by machine decision-making, instant feedback, adaptability, and collaboration between human and machine components. This will involve the elimination of layers or hierarchy, more networked structures, the use of real-time data and analytics to inform decision-making, and the cultivation of a culture of experimentation and continuous learning. The kind of experimentation that Bezos suggests is essential, and the absence of it is "dangerous." By leveraging artificially intelligent systems—nascent implementations of the all-powerful CIC imagined in this chapter—companies can solve communication bottlenecks, eliminate hierarchy, unlock information flows, and create more resilient, innovative, and scalable organizations that continue to deliver growth.

How will we measure the successful implementation of such cybernetic capabilities? When companies implementing them start to break free of West's current scale curves. When they start delivering income growth with a scaling exponent greater than the 0.79 measured by West in his research. Using these laws, we can measure the relative efficiency of human–machine systems when implemented in companies. We know where present systems begin to show sublinear scaling, and we can often identify the cause. Is it our inability to deal with more than a certain number of people? Is it cognitive overload? And if so, what would replacing that human component with a machine component do to the performance and scaling of the overall system?

Just as Bezos knows in his gut that innovation is key, West's work also deals with the necessity of innovation. But West puts this idea in the context of a more formal framework, one that he calls "cycles of singularities." By examining the differences between companies and cities, he finds that the reason cities are able to experience open-ended growth in innovation is that individuals, groups, and pursuits within the city might die, but there is always another idea in the works. Therefore, the city benefits from a kind of Pony Express of innovation. When one horse gets tired, the rider—the city—shifts to the next one. On the other hand, companies are built more tightly around a technology, a market, or even a product, and they often find it hard to switch steeds when the one they are presently riding tires out. West proposes that at this point, innovation is the only force capable of resetting the clock on a system's inevitable decline. Without continuous innovation, all systems eventually stagnate and collapse.

This principle has profound implications for grasping why cybernetic systems are not merely an interesting technological curiosity, but in order to sustain human progress, they are also a necessity. So is the ever-deeper integration of AI into the fabric of our lives. Without it, systems and organizations that exist in a world that moves faster every year will rapidly stagnate and die before they have an opportunity to attain their full potential. It gives new meaning to U2's iconic song "Running to Stand Still."

West's work reveals that the innovation imperative transcends remaining competitive in a dynamic market.

Armed with this view of the world and the future of companies, I asked myself how I would go about building and influencing the companies I am involved with at present. Would I be comfortable outsourcing decisions to AI that I had previously always taken on my own? At SpecFive, a company I founded and where I serve as chairman, I decided to try this out. The company was incorporated in January 2024, and by February 16 of the same year, it had shipped its first product, a mesh-networking device.

How did we decide that we wanted to pursue the decentralized communications market? When I launched SparkCognition back in 2013, it took me about a year of research to fully process all that was going on in the industry. I clearly remember putting together various matrices comparing other companies to what I was thinking of building. I looked at various markets where the company might apply its technology and used a points-based system to determine that industrial AI—an industry that didn't really exist under that definition at the time—was the place to focus.

Just over ten years later, we faced a similar decision with SpecFive, but this time we used AI to look at the communications market, identify needs, analyze user feedback, and develop a series of specific focus areas. Once we picked one, we then honed in on collecting a vast amount of content relevant to that niche and identified more clear and actionable trends and product opportunities. And from that point on, we just kept going. Very soon our AI system produced an entire product road map.

This was a small team of only engineers. Everyone in the company was hands-on, doing things. As of the writing of this paragraph, the company is approaching about fifteen people, but there are still no managers of anything. We're using SpecFive not just as a platform to build great products but also as a case study in building AI-first, AI-inside companies. We're applying AI to finance and modeling future cash flows. We're applying AI to product management and marketing. And we're applying AI to building the products themselves.

The tools we develop in the process of building up the company will be a huge differentiator for us going forward. Perhaps SpecFive or one of the other companies we're building now will become NeuralTech. As a many-time CEO, I have no problem with AI taking over as soon as it can show me that it works.

Our future will require the development of new forms of governance, education, and social safety nets to ensure that the benefits of technological progress are widely shared and that no one is left behind—but at the same time, we must be careful not to abuse regulation and prevent technology from delivering the immense benefits it can readily deliver.

CHAPTER 4

A WORLD OF NEOMS

N eom represents an unprecedented undertaking in its own right, yet its true significance may lie in the ripple effects it could produce across the globe. If this bold experiment in cybernetic urbanism proves successful, it will in all likelihood serve as a blueprint and inspiration for a new wave of urban transformation. In redefining the city, Neom has the potential to set the stage for future human civilization—one defined by the harmonious integration of technology and human will to produce an environment for human progress.

The cybernetic ideas behind Neom can catch on, first in new cities and then for today's metropolises. Saudi Arabia alone has plans to build many new cities, including Qiddiya, a sports-focused city; Amaala, a city that is being imagined as the "Riviera of the Middle East"; Diriyah Gate, a cultural- and tourism-focused city; King Abdullah Economic City; and several others. The United Arab Emirates is also building high-tech extensions to its existing cities and creating entirely new cities from scratch. Masdar has been under construction for more than eighteen years and is aimed at delivering a zero-carbon city. Dubai South, Sharjah Waterfront City, and District 2020 are just a few examples of the new urban environments coming up in the UAE.

The rapidly advancing gulf economies of the Middle East, China, South Korea, and Singapore are all examples of places where developments can occur quickly and at scale. If the cybernetic-city model begins to catch on, these are the places where we will see changes unfolding first. As the principles pioneered and honed in Neom are scaled and replicated across cities and nations, a new era of urban living will be ushered, one that transcends the limitations and inefficiencies of our current models.

To understand the significant impact this can have on demographics, perceptions, and even soft power, take a trip back to the 1980s and 1990s. China had not yet started to develop at a frenetic pace; Dubai and the rest of the Emirates were sleepy coastal settlements; Saudi Arabia was rich, but inward looking and extremely conservative; and more than half of Europe was just emerging from the Cold War. America was the world's technological and cultural capital because it attracted the best and brightest minds from all over. Its military and economic power were unparalleled. Hollywood was light-years ahead of any other movie industry in the world. American music, TV shows, comics, toys . . . they were all leading by a country mile. The United States was simply the coolest country on Earth because it was so much more advanced than other nations and represented the aspirations of many.

Today, things are quite different. Middle powers are emerging in Asia, the Middle East, and Europe. China is the world's largest producer of virtually everything. Its economy has grown by leaps and bounds, and it is the aspirational focus for much of the world. Gleaming Dubai, Doha, and other centers in the GCC (Gulf Cooperation Council) nations present a model of technological sophistication, top-end infrastructure, and modernity that is hard for US cities to match.

Now, in this changing milieu, let's introduce the cybernetic city, an environment so far ahead of anything we're used to that being in a place like that will be a whole different level of "cool." Take the American advantage of the eighties and nineties, and multiply it thrice over. That's what living in cybernetic cities will seem like to those who are on the "outside." The smartest and most capable will want to flock to these oases of innovation: the interconnected nodes in a global network of commerce and economic power.

But it's not just the high life or economic and cerebral pursuits that such cities can deliver. As humanity faces climate change, one of the most significant challenges of our time, ecological and environmental best practices developed in Neom could rapidly disseminate and cross-pollinate across borders, sparking a virtuous cycle of progress. The potential impacts of such a global transformation are staggering. From mitigating the existential threats posed by climate change and resource depletion to addressing urgent challenges in areas like public health, education, and economic inequality, the power of cybernetic urbanism could serve as a force multiplier for humanity's collective efforts.

By learning from the Neom experiment, cities of the future could serve as hubs for carbon-capture and ecological-restoration technologies, pioneering new methods for reversing the damage of industrialization and paving the way toward a sustainable civilization that can regenerate what it consumes. Or envision urban centers that leverage AI and personalized learning to unlock the full potential of every child, regardless of circumstances or background, creating a truly equitable foundation that benefits humanity.

The possibilities are vast indeed.

Cybernetic cities will perhaps not only individually empower their residents and provide them with better economic opportunities, but they might also be the first step in mankind getting past the extractive, exploitative ways of development that have defined modern industrial civilization. We can make these cities horrific models of surveillance capitalism or embrace a more holistic, regenerative worldview rooted in principles of symbiosis, stewardship, and reverence for the interconnectedness of all life.

THE FALLACY OF SUNK COSTS?

Let's balance the optimism with a little bit of brass-tacks mathematics. Neom will cost at least $500 billion to build, and, fully realized, it may cost twice that. Only 28 of the 177 countries on Earth even have a gross domestic product that large! So to expect any other country to be able to afford a project as massively expensive as Neom is, in a word, unrealistic. Does that bit of accounting spell the end of our dream of cybernetic cities across the world? Not in the least. Neom will be a lab. A factory. A place where ideas will be tested. And when you test an idea for the first time, you might spend ten or a hundred times more than you will once you learn exactly how to execute it. On behalf of all humanity, Neom represents a seed investment in the future.

Indeed, even if it might seem to fail, Neom could still pay off. Look no further than the example of Iridium, Motorola's ambitious foray into satellite communications. This was Starlink before Starlink was even a glimmer in Elon Musk's eye. All the way back in the mid-1980s, Motorola imagined a future with global mobile connectivity enabled via satellites. Taking its name from the chemical element with the atomic number of 77, Iridium originally planned to have seventy-seven satellites orbiting Earth, providing seamless communication virtually anywhere on the globe. Although the number of satellites dropped from seventy-seven to sixty-six, the project name remained. Unfortunately for Motorola, the Iridium project initially turned out to be a colossal misstep, characterized by massive sunk costs and facing technological obsolescence upon its debut. In order to build and launch the system, Motorola had invested a staggering $5 billion to deploy the constellation of sixty-six low-Earth-orbit satellites. These satellites were designed to provide global communication coverage to subscribers anywhere on the planet. By the time Iridium became operational in 1998, advancements in cellular technology had already outpaced it, rendering the service too expensive and the bulky, first-generation sat-phone equipment too cumbersome for the average consumer. With service charges ranging from $6 to $30 per minute and phones priced at $3,000, the Iridium business model fell flat in markets where cell phones provided a cheaper and more practical solution. Just a few months before the dot-com bust, Iridium declared bankruptcy in August 1999, becoming one of the most infamous failures in the history of telecommunications.

But the story didn't end here.

Iridium would have a second coming under the leadership of Dan Colussy, a veteran aviation executive with a track record of transforming businesses. Before making a bid for Iridium, Colussy had successfully repositioned UNC Corporation from a struggling nuclear-power-plant builder into a profitable aviation-services provider, which he sold to General Electric for $725 million in 1997. Coming out of retirement to tackle the Iridium challenge, Colussy acquired Iridium's assets during the bankruptcy proceedings for a mere $25 million—a paltry fraction of the original cost. This dramatically lower cost basis allowed for a radical restructuring of the company's financials and operations. Colussy implemented aggressive cost-cutting measures, reducing monthly operational costs from $45 million under Motorola to $3.5 million and minimizing the workforce to essential staff only. This restructuring allowed Iridium to begin to operate viably with significantly fewer customers than it had required under Motorola.

Under new management, Iridium found a niche providing indispensable communication services to specialized sectors such as defense, maritime, and remote operations—areas where conventional cellular networks could not reach. And then a bit of luck came Iridium's way. The company became particularly vital for the US military, which used Iridium's services extensively during its operations in Iraq and Afghanistan. By 2019, Iridium had completed the upgrade of its satellite network with the Iridium NEXT constellation, enhancing its data-transmission capabilities and expanding its service offerings. Iridium's rebirth demonstrated the practical utility of the satellite network and also secured financial success: the company reported sustained revenue growth and expanded its customer base across multiple industry verticals.

It wasn't that satellite connectivity was a bad idea. It was just that the first time it was built, it was way too expensive to be a viable business. Yet once the burden of the $5 billion initial investment was shed, the company could emerge from being a financial disaster to becoming a niche market leader. And so with Neom: just because building the world's first cybernetic city will be expensive doesn't mean that there's something fundamentally flawed about the concept or that it won't eventually catch on.

Almost irrespective of what happens to Neom, we can—and should—abstract out the hundreds of billions of dollars in construction costs for new buildings and focus instead on the first principles of a cybernetic city, asking how to bring those to the cities we already have.

CYBERNETIC CITIES, KNOWLEDGE, AND POWER

Neom's pioneering vision of a cybernetic city seamlessly integrating artificial intelligence, renewable energy, and cutting-edge technology prompts us to think of a world where such cities, villages, and towns become the global norm. Extrapolating this model internationally would mark an urban epoch—a paradigm shift redefining how we conceive and inhabit cities. These cybernetic metropolises would exemplify unprecedented efficiency, sustainability, and synergy between technology and the environment, radically transforming our understanding of urban living.

The implications of this are not just technological but deeply economic and social as well. To understand its full scope, it's worth exploring the insights of an influential thinker on technology and economics: George Gilder.

Gilder was born in 1939 and, after a comfortable upbringing in New York and Massachusetts, obtained a degree from Harvard. Early in his career he served as a speechwriter for several notable leaders, including Nelson Rockefeller, George W. Romney, and Richard Nixon. As an economist, author, and venture capitalist, Gilder has long been at the forefront of exploring the intersection of information technology and economic growth. In his book *Knowledge and Power* he offers us a reframing of our understanding of capitalism and the driving forces behind economic progress.

Through the lens of Gilder's book we can peer ahead into the economic implications of scaling Neom's vision worldwide. Gilder posits that capitalism, at its core, is an information system rather than merely an incentive system built around labor and capital. His reframing highlights the central role of knowledge, innovation, and the technology that enables their propagation in driving economic expansion.

Viewed in this way, a global network of cybernetic cities, villages, and communities like Neom represents the apex of an information-based economic engine. Their integration of artificial intelligence, data systems, and cutting-edge digital infrastructure optimizes the creation, flow, and productive application of information and knowledge across every facet of urban operations.

From intelligently automated supply chains and resource allocation to AI-assisted design and predictive modeling, these cities could become concentrated crucibles for transforming raw information into revolutionary products, services, and solutions.

These cities would produce more data, they would consume more data, and their AI models would learn at a faster rate, not only by watching their immediate environment but also by observing the society of other AI models.

As powerhouses concentrating human capital and next-generation informational industries, these cities may become irresistible magnets for investment and top talent worldwide. This could intensify the economic divide between nations excelling in this arena and those left behind. Conversely, pioneering technologies incubated in these urban crucibles could eventually pervade outward, sparking new industries and economic prospects in underserved regions. The productivity gains from AI integration could liberate human resources for more socially constructive endeavors like education and health care.

Yet Gilder's work also highlights the inherent unpredictability at the heart of informational economies and innovative expansion. Drawing from Claude Shannon's information theory, he frames entrepreneurial activities as the engine creating new economic value by introducing "surprise" into the system—the entropic sparks that ignite new possibilities. If entrepreneurship is the signal of capitalism, the mundane humdrum of big-company operations is the indecipherable noise. The signal, the information, comes from the "surprises." If you take this view, cybernetic cities become urban laboratories optimized for fomenting creative entrepreneurialism and seemingly chaotic experimentation that generate game-changing "surprises."

The staggering complexity and capital requirements for building a single cybernetic city, let alone a worldwide network, present an unprecedented logistical and collaborative challenge across sectors. The first principles of cybernetic cities will have to be distilled and repackaged far more economically via AI applications, mesh networks, sensors, and democratized knowledge. Integrated governance and interoperability protocols enabling cross-city AI ecosystems, data flow, and shared best practices will need to be meticulously constructed.

If we manage to pull it off, this global vision of the rebirth of our cities as living, thinking cybernetic structures actively collaborating with us to deliver a better life can bring about remarkable humanitarian and planetary progress. Harnessed optimally, cybernetic cities' capabilities powered by intelligent information systems could help bend the curve on urgent existential risks like climate change, automation anxiety, pandemics, and societal fragmentation. A meshed global intelligence supercharging sustainability, health-care access, and cooperative problem-solving could catalyze crucial innovations addressing humanity's greatest threats. And the connective fabric woven among cybernetic cities may inspire deeper transcultural understanding and collective planetary custodianship.

One of the greatest opportunities may lie in the potential for cybernetic cities to rebalance the relationship between centralized power structures and distributed flows of information, aligning with Gilder's vision. Historically, he argues, the centralization of governance and economic control has stifled innovation by concentrating authority in the hands of those insulated from the decentralized wellsprings of new knowledge. Think back to the exchange between Orrin Hatch and Mark Zuckerberg.

But in the cybernetic-city model, with its open digital architectures and integrated artificial intelligence, the diffusion of information and decision-making could become radically democratized. This democratization of information flow, when coupled with Neom's emphasis on individual data ownership and privacy rights, could give rise to entirely new models of participatory economics. Citizens could monetize their data streams, capitalizing on the economic value that their information generates for AI systems. Markets for information-based services and products could thrive, turbocharging the innovation that Gilder celebrates.

Moreover, by encoding ethical principles around transparency, equity, and individual empowerment into the core governing code and the AI systems orchestrating the cybercities, we could construct urban environments optimized for human flourishing from first principles. Rather than legacies of historical injustice and inequality baked into legacy urban infrastructure, cybernetic cities could embody a new social contract enabling the full creative expression of what unified human–machine intelligence can achieve.

For ten years, that's what I did at SparkCognition, the AI company I founded in 2013. Our work across various industries provided me with a glimpse into how the transformative potential of cybernetic urbanism can be unlocked in existing cities without the need for trillions in infrastructure investment.

At its core, the promise of a cybernetic city lies in the seamless integration of artificial intelligence into every aspect of urban life. It's about leveraging data and machine learning to optimize city services, enhance public safety, boost sustainability, and elevate the quality of life for residents. And as our work demonstrated, much of this can be achieved by retrofitting existing infrastructure with intelligent software and low-cost sensors.

Take public safety, for instance. SparkCognition's Visual AI Advisor is a software application that implements advanced computer vision and can identify patterns and actions that are unfolding across multiple frames of video. It can transform existing, inexpensive, and entirely ordinary CCTV cameras into smart safety systems that can detect threats in real time. By analyzing video feeds with advanced vision algorithms, it can spot when an elderly person encounters a fall, suspicious activities, unauthorized access, doors left ajar, and even weapons, alerting authorities immediately. Imagine this technology deployed at scale across a city, turning the existing CCTV network into an always-on, AI-powered safety net.

Or consider the challenge of sustainability. Cities account for more than 60 percent of global energy consumption and 70 percent of carbon emissions. Tackling this requires a massive shift toward renewable energy and optimized resource management. That's precisely what SparkCognition's renewable-energy platforms have been focused on enabling. The company has worked with very large renewable energy businesses such as AEP Renewables, Primergy, and European renewables giant Orsted. By analyzing data from solar panels, wind turbines, and smart grids, this software predicts failures, optimizes performance, and maximizes clean-energy production. Deployed across a city's energy infrastructure, such systems could accelerate the transition to net-zero emissions. Compared to the cost of building an entirely new city, the investment is a rounding error. Applied on a larger scale, these technologies could help cities phase out fossil fuels without compromising energy security.

But perhaps the most exciting aspect is how AI can enrich the daily lives of city dwellers. Imagine your daily commute optimized in real time based on live traffic data, using not just the fastest route but also factoring in your preferences for scenic views or avoiding noisy construction zones. Now take it a step further—your AI assistant notices that you have a slightly elevated blood pressure from the last few readings it analyzed via your wearable device. Instead of rushing through the fastest, most direct route, it adjusts the drive, considering your schedule and the fact that you have a few extra minutes. It chooses a path that winds through green spaces, calmer residential streets, or even areas with more natural scenery so that you not only arrive on time but also feel more relaxed and balanced than if you had taken a more stressful, congested road.

Picture your health-care system not just reacting to your symptoms but also predicting them based on your medical history, lifestyle, and environmental factors, offering personalized advice or interventions before issues arise. Your education could be customized to your learning style, using AI-powered tools that adjust the material's complexity based on how quickly you're grasping new concepts or even providing content in more interactive formats if you're a more visual learner.

Then there's the mundane but necessary aspects of life, like paying utility bills or managing city-related permits and applications. With the help of AI and smart contracts, these tasks can be handled automatically, without human intervention. For instance, imagine you need to have a dead tree removed from your property. Instead of filling out forms, waiting for approval, and manually hiring a vendor, you simply take a picture of the tree. A smart-contract oracle analyzes the image using computer vision, confirming that the tree is indeed dead. It then cross-references the image with records from a city survey drone to ensure that the tree in question matches the one on city file—right down to its species and position on your property.

Next, the system checks to see if you've already contracted with a tree-removal vendor that is certified by the city and is currently in good standing. How does it know the vendor's status? It knows because the city has cryptographically signed the certificate of good standing for this vendor and placed it on an immutable, unhackable blockchain. If everything checks out, the smart contract automatically approves the tree removal and schedules both the service and a follow-up inspection by a city-run drone. This post facto drone inspection ensures that the removal was completed as reported and that all city regulations were followed, closing the loop on compliance without the need for human oversight.

These are not far-fetched scenarios; they are the logical extensions of the AI applications we're developing today. Intelligent systems will streamline not just tree removal but all sorts of city-related interactions, from construction permits to waste management. These advances will free up valuable time for citizens and city officials alike, ensuring faster responses, lower costs, and more transparency in everyday civic tasks. AI isn't just about making complex tasks easier—it's also about simplifying and optimizing the very fabric of urban life, even down to something as personal as ensuring that you arrive at your destination in a state of calm and well-being.

As I reflect on the work we've done already, I can't help but feel optimistic about the future. I know that software and inexpensive sensors and actuators connected over wireless networks can go a long way in improving lives. I know this because I've seen firsthand how AI can unlock immense value from existing assets and systems. We've demonstrated that the building blocks of cybernetic urbanism—the sensor networks, the data analytics, the predictive models, the autonomous systems—are not the stuff of science fiction, but the tools of today. I *know* that we can build cybernetic cities in a cost-efficient way.

So while a city like Neom may seem like a distant dream, the reality is that we already have the technology to start building the cybernetic cities of tomorrow right here in the urban fabric of today. By focusing on the first principles—the core capabilities and benefits that define a cybernetic city—we can chart a path to a future where the transformative power of AI is not confined to greenfield projects but is woven into the very fabric of our existing communities.

LIVING IN A WORLD OF CYBERNETIC CITIES

Imagine your logistics networks seamlessly connected and optimized via data sharing between cities. They span autonomous vehicles, rail transport, buses, urban aerial-mobility systems (flying cars!), and airlines. As you communicate your decision to move from point A to point B, your personal AI obtains all manner of data, including micro-weather and up-to-the-minute traffic information from drone networks and mesh networks, and optimizes your journey. At every step, the AI working with the infrastructure of the smart-transportation stations does the work for you. A dozen technologies come together to enable experiences for you, dynamically constructed in nearly real time to suit your needs in the moment: the augmented-reality glasses you might wear to illuminate directions, arrival information, and nudges on where to go and where to turn; the NFC (near field communications) and Bluetooth tags you carry or come into contact with that ensure you're where you're supposed to be even when you're in a subterranean hyperloop tunnel unconnected with GPS; and the transaction-fee-free blockchain payment networks that work safely, freely, and quickly while being open to all and controlled by no one corporation.

As you arrive at the building where your meeting is scheduled, the building elevators know who you are and are already aware of which floor you need to visit. The building shares data with your AI to let you know where the lobby and the restrooms are, and automatically lets your hosts know that you have arrived. There are no last-minute logistics making you nervous. Everything is taking care of itself.

As you sit down and start discussing the licensing of your technology, developed in a cybernetic city in Qatar and being licensed to a company operating in Munich, the latest intellectual-property laws are already being accessed, and your personal AI is continually checking with Munich's regulatory agency APIs to ensure that what you are proposing is doable. You don't need to ask. The AI and the city's regulatory APIs are doing the job for you.

As you arrive at a deal, you wonder how you will deliver some of the licensed technology to the partners you are meeting. Who will warehouse your products? How will they be produced? What type of duty, demurrage, and clearing fees will you be on the hook for? What capacity will you be able to deliver? Luckily, the city is aware of all the businesses that operate within it and also understands how to query their systems for available capacity information, rate quotations, and timeline estimates. On the side of these manufacturing companies, more machines are doing the talking because they know what they are building and how much they can build. They can quite accurately estimate everything they are being asked to provide. Your AI—your extended cybernetic self—is also figuring out warehouse and distribution opportunities. The digital infrastructure of sensors, AI, and data in cybernetic cities has transformed everything into a digital asset, even storage space. Storage units know whether they are empty or full. Warehouses can predict their availability and determine if the environmental conditions they offer are suitable for your product.

It's not just a car that drives itself autonomously in a cybernetic city; transactions and business undertakings that would take an army of people months to figure out are happening on the fly, being constructed in real time driven by a live dialogue.

And that's just a small vignette of the art of the possible.

Extrapolating Geoffrey West's findings, which are covered in the previous chapter, to a worldwide mesh of cybernetic megacities is staggering. The advanced AI integration and massive size and density could generate self-reinforcing feedback loops that drive innovations at an exponential, rather than linear, rate. Such cities would become concentrated crucibles for radically accelerating technological and social progress across sectors.

The seamless interoperability of companies, infrastructure, and logistics among cybernetic cities could amplify these effects. Instead of innovations developing in isolated silos, a global meta-city network enabling the frictionless flow of data, intellectual property, and human capital could spark compounding synergies and cross-disciplinary technological spillovers at a truly planetary scale. This seamless circulatory system for the world's top minds and innovations could bend the arc of progress like never before, rapidly solving our greatest collective challenges.

For example, breakthroughs in decentralized AI networks, smart-mobility solutions, or renewable infrastructure pioneered in one cybernetic city could rapidly cross-pollinate enhancements and applications in others. And with their AI backbone, these cities could optimize this diffusion, intelligently routing innovations and intellectual resources with unprecedented efficiency to hot spots of related activity. The accelerating network effects and superlinear outputs could become self-perpetuating as cities compete in an innovation arms race to remain premier cybernetic hubs.

Why would cities at such scale and interconnectedness even be possible? Because with the augmentation of AI, humans might, for the first time since their origin, begin to extend cognitive constraints such as the Dunbar number, discussed in the next chapter.

DEMOCRATIZING CYBERNETIC URBANISM

With the right approach and the right tools, the benefits of this new cybernetic urban paradigm can be extended to communities and contexts around the world, including some of the most resource-constrained and underserved areas.

A powerful example of this potential can be found in Project TechHub, an initiative spearheaded by my wife, Zaib, and me in partnership with the International Board of Books for the Young (IBBY) and the long-serving Pakistani nonprofit Alif Laila Society. The project aims to bridge the digital divide in rural Pakistan by distributing laptops, tablets, and robotics kits to schools in remote areas, starting with Dera Allahyar in Baluchistan and now extending to the province of Sindh.

At first glance, the connection between a local educational-technology initiative and the grand vision of the cybernetic city may not be immediately apparent. But look closer, and you'll see that Project TechHub embodies many of the core principles and strategies that will be essential for democratizing access to the benefits of cybernetic urbanism.

At its heart, Project TechHub is about empowerment through technology. It recognizes that in an increasingly digital world, access to computing devices and the skills to use them is not a luxury but a fundamental necessity for full participation in society and the economy. By putting these tools in the hands of young girls and boys in underserved communities, the project is not just enhancing their educational opportunities in the short term. It is laying the foundation for a future in which these children can grow up to be the scientists, engineers, business owners, and leaders who will drive innovation and positive change in their communities and beyond. The technology will change them, and they will use the technology to change the world.

Through hands-on training and curriculum development, Project TechHub is empowering these children to use technology and also to create and manage it. They're learning how to set up and maintain mesh networks using low-cost, open-source solutions like those being developed by my company SpecFive. These decentralized networks allow for resilient, community-owned connectivity enabling students and local stakeholders to share resources, knowledge, and data without relying on traditional infrastructure.

Moreover, we're rolling out programs introducing students to the world of AI and the internet of things (IoT), teaching them how to leverage the computing power of their devices to run machine-learning models and collect data from sensors deployed in their environment. Imagine a future in which these students can use their skills to monitor local agricultural conditions, optimize crop yields, and share insights with farmers across the region—all through a network they build and manage themselves.

This bottom-up approach to technology deployment and capacity building is key to democratizing the benefits of cybernetic urbanism. By empowering communities to take ownership of their digital infrastructure and leverage it to address local challenges, we can create a more inclusive and equitable vision of the smart city—one that extends beyond the confines of high-tech enclaves like Neom. And we can invite even the least privileged among us to extend themselves cybernetically by integrating themselves with technology.

Of course, realizing this vision will require more than just technological innovation. It will demand new models of collaboration with remote government-managed schools of very little means, investment, and governance to ensure that these community-driven networks are sustainable, secure, and accountable. But as initiatives like Project TechHub demonstrate, the seeds of this transformative potential are already being sown.

Access to devices is only part of the equation. To truly unlock the potential of these tools, they need to be connected to one another and to the wider world of information and opportunity. That's where the concept of mesh networking comes in—and it's an area where SpecFive is working to drive transformative change.

Mesh networks are decentralized, self-organizing networks in which each node (or transmitter/receiver device) can relay data for all other nodes, allowing for the creation of large-scale, resilient networks without the need for centralized infrastruc-ture. This makes them particularly well-suited for contexts where traditional connectivity solutions are unavailable, unreliable, or unaffordable—such as in rural or remote areas, or in the after-math of disasters that have disrupted conventional communication systems.

SpecFive is already building software and hardware solutions to make mesh networking an instant-on experience for users, supporting several existing protocols such as the open-source Meshtastic and our own in-development HyperMesh protocol. The company's goal is to enable the creation of the world's largest mesh network, providing a low-cost, secure, and scalable connectivity solution for billions of IoT devices and users around the world.

The potential applications of this technology are substan-tial. The laptops and tablets distributed through Project Tech-Hub won't just be stand-alone devices but nodes in a resilient, community-owned mesh network that eventually covers the entire region. Suddenly, these tools are not just portals to preloaded edu-cational content but also gateways to a universe of knowledge, communication, and collaboration.

Through the mesh network, students and teachers across vil-lages could share resources and ideas, collaborate on projects, and access online-learning platforms and expert instruction from around the world, all without paying a cell-phone charge. Local farmers could use the network to monitor soil conditions, track weather patterns, and access real-time market information to opti-mize their crops and increase their incomes. Health-care workers could use the network to coordinate the delivery of supplies and services, access telemedicine support, and track public-health data to prevent and respond to outbreaks. And they can do all of this without paying cell-phone companies monthly fees. In developing countries, almost any amount that needs to be paid on a monthly basis can pose issues of affordability, especially when enabling a large number of devices such as agricultural sensors and controls.

In essence, the combination of device access and mesh networking would bring many of the key benefits of cybernetic urbanism— the real-time data flows, the optimized resource management, the enhanced service delivery—to some of the most underserved and isolated communities on the planet. And it would do so not through a top-down, centralized model of technology deployment but through a bottom-up, community-driven approach that empowers local actors to shape and own their digital infrastructure.

This is just one example of how the principles of cybernetic urbanism can be translated and adapted to diverse contexts around the world. And it highlights the immense potential for projects like TechHub and technologies like mesh networking to serve as catalysts for a more inclusive and equitable vision of the cybernetic city—one in which the benefits of digital transformation are not confined to ultramodern enclaves like Neom but are accessible to communities and individuals everywhere.

A child in a remote village in Pakistan should be able to access the same educational resources and opportunities as her peers in the most advanced urban centers, and actively contribute her own knowledge, creativity, and perspective to a global community of learners and innovators. A farmer in Cameroon should be able to monitor and optimize his own crops, and then share his insights and experiences with a worldwide network of agricultural practitioners and researchers, cocreating new solutions to the pressing challenges of food security and climate resilience.

This is the true promise of a democratized approach to cybernetic urbanism—one in which the tools and benefits of the digital revolution are not just accessible but are actively shaped and wielded by communities everywhere. For me, the true measure of the cybernetic city will not be in the brilliance of its mass transit or the gleam of its towers, but in the opportunities it creates, the lives it enriches, and the dreams it sets in motion for billions of people around the world. And that is a future worth building—one mesh node, one tablet, one curious mind at a time.

Through Gilder's informational lens, such a global cybernetic-city network—if we can build it—represents the apotheosis of an economic system built to maximize the generation of knowledge and its productive application. Through their digital infrastructure, data-sharing protocols, and the human–AI synergies they enable, these networks could remove long-standing bottlenecks to the kind of free flow of information that constrains growth. The economic value created as a result will be staggering. And as machine intelligence expands our capacity to generate and capitalize on the informational "surprises"—the signals—that drive breakthrough innovations, this economic value will only grow.

Yet as Gilder discusses, the diffusion of information and the decentralization of innovative activities tend to naturally distribute economic power and destabilize the status quo. And you can freely read "status quo" here to mean entrenched hierarchies—politicians, bureaucrats, large companies. A world-spanning cybernetic urban network structured around an open-information architecture and AI-enabled participatory governance could fundamentally reshape local and global politics. If nothing else, cybernetic cities will unleash a new renaissance of creative entrepreneurialism and cooperative problem-solving at a planetary scale.

What do we have to overcome to achieve this vision of an interconnected world of humans and technology helping one another build and thrive? For this, we need to consider the skills we need to develop as humans. And then, in Chapter 7, turn to a data-driven science of civilizational cycles called cliodynamics. Let's start with the humans.

CHAPTER 5

HUMAN AUGMENTATION

Ava sits in her living room, her eyes closed, her mind focused. She's not meditating or daydreaming but rather engaging in a complex dance of thought and intention, a silent conversation with the sleek device resting gently on her head. This is her brain–computer interface (BCI), and it's about to take her on a journey beyond the boundaries of her physical self.

With a subtle mental command, Ava activates her BCI's virtual-reality mode. Instantly, her living room dissolves, replaced by a stunningly realistic digital landscape. She finds herself standing on a mountaintop, gazing out over a vast, vibrant valley. The air is crisp and clear, the sun warm on her skin. She tilts her head back as the gentle breeze blows through her hair. A flock of birds lifts into the air, calling out as they ascend. There are no actual birds in the living room, but her brain registers them as real through the neural link. For all intents and purposes, Ava is there on the mountaintop, her senses fully immersed in this digital world.

But this is no mere entertainment experience. Ava is studying to be an ecologist. Traveling the globe for lectures or exams isn't feasible, but the BCI reduces the need for such travel. Her living room is her classroom. As Ava explores this virtual environment, her BCI is hard at work, monitoring her neural activity, tracking her eye movements and pupil dilation, and even sensing subtle changes in her facial expression and skin conductance. This rich stream of biometric data is fed into a sophisticated AI system that adapts and optimizes the experience in real time.

This simulation is based on real-world data from global ecological sites. It's also set up as her final exam. Ava must identify various sources of ecological distress, facing puzzles integrated into the model. But she has prepared. When she encounters the first challenge—a fungal growth on tree roots—Ava identifies the cause with ease. The BCI registers her confidence and adjusts the difficulty to continue testing her. Every test is personalized, designed specifically for her thinking patterns. No two students will face the same exam.

As Ava moves deeper into the valley, the puzzles grow more difficult. When she faces a particularly challenging problem, frustration mounts as time runs out. The BCI notices and subtly adjusts, offering a hint that guides her in the right direction. Her curiosity spikes, frustration fades, and she solves the puzzle just as the exam ends.

The goal isn't to fail students but to explore their knowledge and strengths. The BCI notes when Ava's excitement or curiosity grows, generating new challenges that keep her engaged. This is a deeply personalized, adaptive experience that feels less like interacting with a machine and more like an extension of Ava's own mind.

CYBERNETIC COEVOLUTION

A story like Ava's may seem impossible today. How could we grow accustomed to such deep intrusions in our minds and lives? Yet this has always been the case with humans and technology. The line we often draw between ourselves and our tools is artificial. In truth, humans do not merely adapt to technology; we coevolve with it. This reciprocal relationship has shaped who we are from the very beginning.

Take smartphones, for example. A study published in *Current Biology* and reported on by *Wired UK* found that extensive smartphone use has reshaped the sensory relationship between our brains and thumbs. Our bodies are changing in subtle ways to accommodate these devices. The phenomenon known as "Swiper's thumb"—where the dominant thumb increases in size by up to 15 percent as a result of frequent swiping—is a testament to this ongoing biological adaptation.

As Nina Bibby of O2 noted, "It's difficult to tell where our hands stop and the handset starts." This blurring of boundaries is a hallmark of our coevolution with technology. Smartphones, once thought of as external tools, have become extensions of our physical and mental selves. Now imagine what happens when we can communicate directly with our personal AI at the speed of thought.

Companies like Neuralink, Paradromics, Synchron, and Braingate are developing technologies that will make this possible. The convergence of brain–computer interfaces and AI represents the most direct way that consciousness can influence technology—and vice versa. The idea of "prompt engineering" as we know it today will evolve into thought engineering, where desires are directly translated into digital or mechanical action.

FUSION SKILLS OR COEVOLUTION?

Paul R. Daugherty and H. James Wilson, in their book *Human + Machine*, argue for the development of "fusion skills": the ability for humans to collaborate effectively with AI systems. Although this is a valuable perspective, their view risks oversimplifying the relationship between humans and machines. They treat AI as a tool to augment human abilities rather than recognizing the deeper reality of coevolution.

Daugherty and Wilson emphasize "bot-based empowerment" and "rehumanizing time," arguing that AI will take over mundane tasks, freeing humans for creative pursuits. But this perspective neglects the extent to which humans and AI are already merging. The focus should not be solely on using AI to "extend" ourselves but on the fact that AI and humans are growing together— integrating in ways that reshape our biological and cognitive realities.

For example, the idea that AI will simply give us more time for creative pursuits assumes that humans and AI will remain separate entities, with AI assisting as needed. But the reality of brain–computer interfaces, neural networks, and personal AI suggests something far more profound: a symbiosis in which AI becomes part of how we think, feel, and act.

Daugherty and Wilson's fusion skills fail to fully appreciate this. They treat AI as an external aid rather than a coevolutionary partner. In contrast, cybernetic coevolution views the merging of human and AI intelligence as inevitable. Consider the growing role of personal AI systems in managing our lives, anticipating our needs, and even assisting in decision-making. These systems are not just tools; they are extensions of ourselves.

What will it take to thrive in this cybernetic future? Beyond Daugherty and Wilson's fusion skills, I would suggest we embrace a mindset that sees technology as a partner in our evolution. This means cultivating skills like adaptability, continuous learning, and ethical AI navigation. But more than that, it requires us to rethink what it means to be human in an age of AI.

Here, the ancient Ship of Theseus paradox comes to mind: If we replace one plank of the ship, it remains the same ship. But if we replace every plank, is it still the same ship? Consider this in the context of human augmentation: If we add one bit of technology to ourselves, are we still human? If we add a hundred, are we still human? If most of our abilities are now outsourced to machines connected with us through BCIs, are we still human?

This philosophical question asks us to confront the nature of our identity in a future where the boundaries between human and machine blur. As we incorporate more technology into our biology—enhancing our cognitive and physical abilities—does the essence of what makes us human remain intact?

Taking this further, what is the place of the human in an integrated machine of this type? Is it simply a matter of relative computational scale—our biological brain as one part of a larger machine, dwarfed by the computational power of the AI? Or is it, as I discussed in *The Sentient Machine*, a matter of how neural networks, even biological ones, are initialized? The initialization of a neural network—how it starts and develops its unique biases—gives it perspective. This perspective, this bias, shapes how it uncovers the infinite landscape of ideas.

We as humans are driven to discover that landscape, and this drive is part of our core identity. Speed, in this sense, is an interesting but not particularly important consideration. Whether we are slow human minds or fast machine minds, both are faced with the same infinite landscape to explore. And in that context, speed becomes irrelevant—neither fast minds nor slow minds can ever hope to traverse the entirety of infinity.

What truly matters is how you approach this exploration. Where do you start? How do you go from idea to idea? These questions of discovery and exploration are rooted in how neural networks are biased, how they are initialized, and how this perspective allows each network—human or machine—to perceive things differently. This bias, this initial condition and its subsequent unique evolution, is why any intelligent machine, including the human machine, is important.

Thus, our identity in the future is not defined by how much technology we integrate but by the uniqueness of our perspective, our drive for discovery, and the way our minds—biological or augmented—navigate the vast, uncharted landscape of ideas. This is the true essence of intelligence and the essence of what it means to be human in an era of rapid technological augmentation. To me, the future is not about mastering tools—it's about coevolving with them. BCIs, personal AI, and neural interfaces are not simply extensions of ourselves. They are parts of a larger, integrated system where humans and machines grow together. In this system we don't just adapt to technology; we also cocreate with it, and perhaps we need to see it as part of ourselves, not an artifact apart.

AUGMENTATION IN AVA'S LIFE

In Ava's world, brain–computer interfaces are ubiquitous, used not just for education but in every aspect of life. In schools, BCIs adapt lessons and monitor engagement, optimizing learning for each individual. In the workplace, BCIs allow people to control complex machinery with their thoughts, increasing efficiency and reducing risk. Virtual simulations test design concepts before physical prototypes are built, and failures are corrected with a thought.

Entertainment has also transformed. Couples enjoy virtual date nights, attending the best concerts and plays from the comfort of their homes, even when one of them is away on a business trip. Sports fans can not only watch games—they can also immerse themselves in historical matches or custom-created fantasy leagues based on their preferences. In health care, BCIs allow patients with paralysis to control robotic limbs and communicate with the world through thought alone.

In Ava's world, the integration of AI, BCIs, and personal agents has redefined every aspect of life. From work to play, education to social interaction, the line between human and machine has blurred beyond recognition.

BUILDING AVA'S WORLD

At the heart of these remarkable applications lies a complex array of technologies and techniques that enable the coming together of human thought with digital systems. The first step in any BCI system is signal acquisition. The theory that human thought produces electrical signals was first studied in the last 1800s. In 1875 physician Richard Caton studied electrical phenomena originating in the brains of rabbits and monkeys. In 1890 physiologist Adolf Beck continued this work, studying the electrical activity of rabbits and dogs. When Beck placed electrodes directly on the surface of the brain, he observed fluctuating brain activity. This led to the theory of brain waves.

Signal acquisition involves recording the electrical activity of the brain using a variety of methods. One of the most common techniques for signal acquisition is electroencephalography (EEG), which uses electrodes placed on the scalp to measure the collective activity of millions of neurons. Although EEG had been performed on animals, in 1924 Hans Berger recorded the first EEG on a human, after inventing the device known as the electroencephalogram. EEG is currently used in epilepsy monitoring, where it is considered the gold standard. By monitoring patients, both during seizures and in the time between seizures, treatment options may be identified. It may also help distinguish epileptic seizures from other issues that present the same, such as psychogenic nonepileptic seizures and syncope (fainting). EEG is also being used in the diagnosis and treatment of strokes, brain tumors, inflammation of the brain (encephalitis), and sleep disorders. EEG is noninvasive and relatively inexpensive, but it can provide only a coarse picture of brain activity: the signals must pass through the skull and scalp before reaching the electrodes. Still, for those most reluctant about the idea of BCI and the merging of brain and mind, EEG may be the first, best step.

For more precise measurements, researchers are turning to more invasive techniques like electrocorticography (ECoG), which involves placing electrodes directly on the surface of the brain. ECoG was first pioneered by Wilder Penfield and Herbert Jasper, two neurosurgeons at the Montreal Neurological Institute, in the 1950s. It was part of their surgical treatment, known as the Montreal Procedure, used to treat patients with severe epilepsy. In the procedure, electrodes placed directly on the surface of the brain recorded electrical activity. When zones were identified where epileptic signals originated, these zones were surgically removed and the cortex resectioned, destroying the epileptic brain tissue. Though more invasive than the EEG, ECoG does not penetrate the blood–brain barrier, the border of cells that protects the brain from unwanted or harmful substances. ECoG offers much higher spatial and temporal resolution than EEG, enabling the detection of subtle patterns of activity that might be missed by scalp electrodes. However, it requires surgery to implant the electrodes, which carries risks of infection and tissue damage.

A more precise method of signal acquisition uses microelectrode arrays (MEAs). There are two types of MEAs: implantable and nonimplantable. Implantable intracortical microelectrode arrays are inserted directly into the cortex and can record the activity of individual neurons. Research shows that these devices can be used to help treat a variety of conditions, including depression, epilepsy, and Parkinson's disease. With epilepsy, for example, when seizure signals are detected, the MEAs can then deliver autonomous inhibition signals, preventing the seizure. For those patients with prosthetics, these limbs can be controlled with the mind. With their direct connection to neurons, these microelectrode arrays offer unprecedented insight into the workings of the brain, but they are also the most invasive and carry the highest risks.

Through whatever means is used, once the brain signals have been acquired, they must be processed and decoded in order to extract meaningful information. This is where advanced signal processing and machine-learning algorithms come into play. By training on vast datasets of neural recordings, these algorithms can learn to recognize patterns of activity that correspond to specific thoughts, intentions, or mental states.

One of the most promising approaches in this area is the use of deep learning, a subfield of machine learning that involves training artificial neural networks on massive amounts of data. Deep learning has already revolutionized fields like computer vision and natural language processing, and it is now being applied to the challenge of decoding brain activity.

For example, researchers have used deep learning to decode imagined speech from ECoG recordings, achieving accuracy as high as 92-100 percent. By training a deep neural network on examples of spoken words and their corresponding brain activity patterns, the system learns to map the neural representations of speech to their acoustic counterparts, enabling the synthesis of speech directly from brain signals. Until now, telepathy has been a creation in the science-fiction world, but with this advent the fiction is becoming the science. Will we be looking at a future where audible speech is reduced and telepathic exchanges via the BCI become the norm? As with all science, as the theory becomes the reality, there will be both pros and cons.

Other researchers are using deep learning to develop more naturalistic BCI control schemes, such as those based on motor imagery. By training a neural network to recognize patterns of brain activity that correspond to imagined movements, these systems can enable users to control robotic arms, wheelchairs, or other assistive devices simply by thinking about moving their own limbs.

Beyond the algorithms themselves, BCIs also rely on advanced hardware and materials to interface with the brain. For example, flexible electronics are enabling the development of more comfortable and less invasive ECoG arrays that conform to the curvature of the brain. As new discoveries in materials science are made, more bio-friendly materials will continue to be invented. Optogenetics, a technique that involves genetically modifying neurons to make them sensitive to light, is opening up new possibilities for precise, optical stimulation of neural circuits.

The rapid advancement of BCI technology is being driven by a growing ecosystem of companies, start-ups, and research institutions, each pushing the boundaries of what's possible in its own unique way.

One of the most high-profile players in this space is Neuralink, the brain–computer interface company founded in 2016 by Elon Musk. Neuralink is developing a fully implantable BCI that uses flexible "threads" studded with electrodes to record brain activity.

The company's short-term goal was announced to be treating serious brain diseases, and the long-term goal was human enhancement, also known as transhumanism. Neuralink plans to create a "neural lace" that can be woven into the brain, enabling seamless, high-bandwidth communication between human and machine. As noted by Musk, the idea for "neural lace" came from the ten-book science-fiction series *The Culture* by Scottish author Iain M. Banks. The probes in the device are made of polyimide, a rugged, plastic biocompatible material. As of May 2023, FDA approval for human clinical trials for testing had been received.

Another key player is Kernel, a start-up founded in 2016 by Bryan Johnson. Kernel is developing noninvasive BCI hardware based on advanced optical-imaging techniques. Kernel's Flux device uses optically pumped magnetoencephalography to directly detect the magnetic fields generated by neural activity in the brain, and its Flow device uses near-infrared spectroscopy to detect subtle changes in blood flow and oxygenation. By combining these measurements with sophisticated signal-processing algorithms, Kernel aims to create a high-resolution, real-time picture of brain activity that can be used for a wide range of applications. Sound ID, software designed by Kernel, can identify what song or speech a person is listening to based solely on brain activity and data. But the aims of Kernel are not solely entertainment. In 2019 the company began researching depression, anxiety, and neurological diseases such as Alzheimer's and Parkinson's.

Meanwhile, Paradromics, founded by Matt Angle in Austin, Texas, is taking a different approach, focusing on the development of high-density microelectrode arrays for intracortical recording. The company's flagship product is the Neural Input-Output Bus (NIOB), a fully implantable device that can record from up to 65,536 individual channels. By providing an unprecedented level of resolution and precision, Paradromics aims to enable a new generation of medical devices and therapies for conditions like paralysis, sensory loss, and neurological disorders.

In the realm of noninvasive BCIs, another technology, electromyography (EMG), may prove to be useful. EMG detects the electrical potential of muscle cells in skeletal muscles when the cells are activated, either electrically or neurologically. Companies like CTRL-Labs (now part of Facebook Reality Labs) are developing wristbands that use EMG to detect subtle muscle movements and translate them into digital commands. By measuring the electrical activity of motor neurons in the arm, these devices can enable intuitive, hands-free control of computers, smartphones, and other devices.

Other companies, like Neurable and NextMind, are exploring the use of EEG for hands-free, brain-controlled interfaces. By combining advanced signal processing with machine-learning algorithms, these companies are creating devices that can detect and respond to specific patterns of brain activity, such as those associated with focus, relaxation, or even specific thoughts or intentions.

THE ROAD AHEAD

As these examples illustrate, the BCI landscape is incredibly diverse and rapidly evolving, with new breakthroughs and applications emerging on a seemingly daily basis. But for all the progress that has been made, there are still significant challenges and ethical considerations that must be addressed as the technology moves forward.

One of the biggest challenges is ensuring the safety and long-term stability of invasive BCI devices. How will the brain be accessed when implanting devices? Although the risks associated with brain surgery are well understood, the long-term effects of implanting electrodes or other devices in the brain are still largely unknown. Researchers will need to carefully monitor the safety and efficacy of these devices over years and even decades to ensure that they are not causing unintended harm. Regulations must be defined to outline who is able to perform these implants. Will it be medical personnel only, or will new implants be as easy to obtain and swap out as getting a new cell phone has come to be? When it comes to the age of a patient, what age restrictions will be put in place? Much like a dental implant, will those receiving an invasive BCI need to be over the age of eighteen? Twenty-five? If an implant stops working, will it be repaired and refurbished? Will the BCI become just one more electronic device, or will it be more akin to a pacemaker?

Another key issue is developing BCI systems that are reliable, robust, and easy to use. For BCIs to truly live up to their potential, they will need to work consistently and seamlessly in a wide variety of real-world environments, from the home to the workplace to the hospital. This will require continued advances in signal processing, machine learning, and hardware design, as well as close collaboration among researchers, clinicians, and end users. All cell-phone owners have experienced the frustration of going through a dead zone or having a provider network go down. The frustration stems from our reliance on our phones. With BCIs, there is no reason to think that reliance will be anything less. It will be much, much more. Signals will need to be available at all times, everywhere. Users will demand seamless and fast updates for BCIs. Security on the devices must be solid.

Perhaps the most profound challenge, however, is grappling with the ethical and societal implications of BCI technology. What are the societal effects of who can and who cannot afford a BCI? Will the divisions of class become even more pronounced? Will educational opportunities continue to divide various income levels? As these devices become more powerful and more widely used, they will raise fundamental questions about privacy, autonomy, and the very nature of human identity. Restrictions and regulations must be put in place, not only on what information companies can access but also on how clear companies must be when requesting access to information. A simple "Agree to All" click box may not suffice. Users need to be made aware of what exactly they are giving the machines access to. Targeted ads may be fine, but when those ads begin to reflect our innermost secret desires, problems will arise.

When our thoughts and intentions can be read and influenced by machines, what does that mean for our basic rights and freedoms? When our cognitive abilities are augmented by artificial systems, where do we draw the line between human and machine? And as BCIs become more accessible and affordable, how do we ensure that they are used in ways that benefit all of society rather than exacerbating existing inequalities and power imbalances?

These are complex and deeply philosophical questions that will require ongoing dialogue and debate as the technology continues to evolve. But one thing is clear: as with AI, the future of BCIs is not just about the technology itself but also about the values and principles that guide its development and use.

BCIs are not the limit, either. Human beings themselves will evolve, augmented in many, many ways through technology. When this happens—as it is happening already—the actions they take will be hard to attribute to just their biological mind. Not in the sense that the enhancements they make will control their mind, but in the sense that these enhancements and augmentations will definitely change the decisions made.

For example, at a trivial level, running from point A to point B tired you, so you stopped. But when equipped with an exoskeleton, you keep going, as researchers at the University of California–Berkeley have demonstrated with their BLEEX (Berkeley Lower

Extremity Exoskeleton) project. This wearable robotic device supports the wearer's legs, reducing the metabolic cost of walking by 1-22 percent. It's easy to imagine how such a device could allow a rescue worker to keep going long past the point of exhaustion or enable a soldier to carry far heavier loads over longer distances. The decision to stop or keep going is no longer just a matter of willpower but can be shaped by the technological augmentation of our physical capabilities.

But the implications go far beyond physical enhancement. Cognitive enhancement will arguably have even more profound effects on our decision-making. Do you choose to be a bit more independent, pushing back against social pressure, when your brain is directly connected to an AI adviser whispering in your ear? Do you become more aggressive in negotiations when an AI is feeding you real-time tips on your counterpart's psychological state based on their facial micro-expressions? Do you take more risks when a predictive algorithm shows you that the odds are in your favor?

These scenarios raise profound questions about agency and responsibility in an age of cognitive augmentation. If an AI influences your decision to take a risky bet that ends up causing harm, who is responsible? Is it you, for ultimately making the choice? Is it the AI developers, for creating a system that encourages risk-taking? Is it the regulators, for allowing such systems to be deployed? As our cognitive processes become more entangled with AI systems, these questions of moral and legal accountability will become increasingly complex.

As neurotechnology advances, we may gain the ability to directly modulate our emotions, attention, and memories. A study published in *Current Biology* demonstrated that stimulating the brain with imperceptible electrical currents can boost or diminish a person's understanding of math. Another study showed that stimulating the entorhinal region can enhance memory recall. As these capabilities mature, the decision to study harder, or to remember or forget a painful experience, may become a matter of dialing up or down a neural implant.

What also of Dunbar's number, the cognitive limit on the number of stable social relationships a primate can maintain, and our ability to engage with and hold a number of people close to us in our networks? Dunbar's number is often cited as a key factor limiting the size and complexity of human social groups. But what if we could augment our social cognition?

Research by Dunbar and his colleagues has shown that active social-network size is correlated with the volume of the orbital prefrontal cortex. This suggests that augmenting this brain region could allow individuals to maintain more relationships. If cybernetic enhancements can expand our Dunbar number, it could have profound implications for the scale and density of our social networks. Larger, more interconnected social groups could accelerate innovation and productivity. Cities capped in size by our cognitive limitations could potentially grow even larger and denser.

However, the expansion of our social cognitive abilities could also have downsides. Robin Dunbar himself argued that our cognitive limitations on social-group size may serve an important evolutionary purpose, preventing us from forming superficial relationships at the expense of deep, meaningful ones. If we can maintain thousands of relationships, will they be of the same quality and emotional depth as the handful of close bonds we maintain now? Moreover, the ability to maintain vast social networks could amplify the spread of misinformation and polarization, as we've already seen with the rise of social media. Managing the negative externalities of expanded social cognition will be a key challenge.

The Dunbar number is only one of the in-built limitations of human design that could be overcome. What about augmentation via neural interfaces that can give us unlimited memory by acting as a conduit to data-storage systems, and give us unlimited intelligence by tying us into personal AI and computational tools? And even the ability to play forward thousands of futures mentally, in concert with the mobile, miniaturized computers we carry on us today.

Ava's exam score comes through. She has passed the exam. In addition, because of the feedback received by the BCI, her course schedule for the upcoming semester has been set. It's even more than she hoped for. Courses have been created and personalized for her, as she is certain they have been for other students in the program.

As Ava removes her BCI headset and blinks back into the familiar surroundings of her living room, she reflects on the exam and everything that made it possible. It's not just the virtual adventure and customized course schedule that continues to awe her, but also the wider arc of technological progress. The customizing of her education and career to her personal strengths. She knows that she is living through a pivotal moment in human history—a time when the boundaries between mind and machine are becoming increasingly blurred.

It's a prospect that is both exhilarating and daunting. But as she looks to the future, Ava is filled with hope. She knows that this integration of mind and machine won't always be easy—that there will be missteps and potholes along the way. But she also knows that by working together, by bringing the best minds and most cherished values to bear on these profound questions, a future can be shaped in which the power of technology is harnessed for the good of all.

In this future, BCIs will not be a tool of oppression or control but a means of empowerment and liberation. They will not replace our humanity but will help us to fully realize it—to unlock the vast potential that lies within each of us and to connect with one another and with the world in ways we can now only imagine.

As Ava sets her BCI aside and steps out into the sunlight of the real world, she carries this vision of the future with her. It is a vision of a world where technology serves the needs and aspirations of humanity, where innovation is moderated by wisdom and compassion, and where the boundaries of the possible are forever expanded. It is a vision of the future that we are building together, one thought at a time.

BEYOND AVA'S WORLD

Indeed, the true power of a BCI isn't even what we see in Ava's world, but in what might emerge from it.

Upon the fabric of existence, the most captivating patterns emerge not from individual threads but from the intricate ways they intertwine. This is the essence of emergence.

Emergence occurs when a complex entity exhibits properties that its individual parts do not possess on their own. Such properties arise only through the interactions within the whole system. Emergence is not an abstract idea but a principle observed across the universe, from the quantum realm to the vast networks of human society.

The concept of emergence has been formalized in various fields, including physics, biology, and computer science. In physics the behavior of complex systems, such as phase transitions in materials, can be described using emergent properties that arise from the collective interactions of the system's constituents. Similarly, in biology the self-organization and collective behavior of cells give rise to the emergent properties of tissues, organs, and organisms. The intelligence of an ant colony—its ability to build a home, to defend it, to forage widely and efficiently—is not a property of any individual, but of their collective interactions. So, too, the individual cells of a slime mold, which can disperse and act alone, or unite and act as one, as conditions demand.

The human brain is also an example of emergence. A neuron in isolation is relatively simple, capable of basic electrical and chemical processes. However, when interconnected in the networks of the nervous system, neurons give rise to consciousness, creativity, and the ability to ponder the cosmos itself. The brain's capabilities emerge from the complex interactions within the entire network, not from any single neuron. Researchers at institutions such as the Wu Tsai Neurosciences Institute at Stanford are exploring how linking neuroscience and AI can advance our understanding of these emergent properties in the brain, with the goal of countering diseases and developing AI technologies inspired by human intelligence.

The human brain consists of approximately 86 billion neurons connected to one another via 100 trillion synapses. The complex interplay of these connections gives rise to the emergent properties of cognition, perception, and consciousness. By studying the brain's network dynamics, researchers aim to unravel the mechanisms underlying these emergent phenomena and apply this knowledge to the development of advanced AI systems. Even though there have been large-scale brain-imaging projects such as Henry Markram's "Blue Brain Project," funded by the European Union to the tune of one billion euros, we have yet to see real success in fully understanding these underlying mechanisms.

Emergent behavior extends beyond biological systems, from water's wetness to the physics of a pile of sand and far beyond. And we can look for them in cybernetic systems, too, and consider the concept of superminds—collective intelligences arising from human–machine interactions. Just as the collective behavior of ants in a colony or cells within a slime mold can give rise to complex, intelligent behavior that no single entity could achieve, networks of humans and machines can produce unfathomable outcomes.

In various fields, AI has been integrated into collaborative roles, aiding in tasks that range from composing music with tools like Flow Machines Professional to assisting in health care with systems such as the SAGE patient-management system. These systems demonstrate how AI can enhance human capabilities by taking over repetitive tasks, thus allowing humans to focus on more complex decision-making processes.

The evolution of human–machine collaboration began with simple tools designed to amplify human physical capabilities. Over time, technological advancements have extended to augmenting cognitive functions. Computers, once large and inaccessible, have become personal companions integrated into daily life. HP's AI-enhanced call centers exemplify this partnership's evolution, showcasing how AI handles routine inquiries and routes complex cases to human agents. This allows human agents to concentrate on more nuanced issues, enhancing overall service quality and creativity in problem-solving. HP's initiative is part of a broader trend of leveraging AI to transform the workplace, as evident from the showcasing of AI-powered solutions at the HP Amplify Partner Conference in 2024.

Implementing AI in contact centers addresses cost reductions while maintaining 24/7 availability. Advanced virtual agents and predictive analytics enabled by AI can help businesses improve operational efficiency, lower costs, and exceed customer expectations. These AI systems often make predictions about humans— which agent is best positioned to handle a query, how systems should prod operators, and how to contextualize the customer's questions with information pre-populated on the agent's screen. They also provide real-time guidance to the agents, suggesting relevant information and solutions based on the customer's history and the context of the conversation.

In essence, the AI is "governing" the human agent's actions. The combined system fuses together into an embryonic supermind capable of delivering far more than either human or machine alone. AI shapes human responses, resulting in a cybernetic response that is neither purely human nor purely artificial. Conversational AI platforms like Google's Dialogflow enable this by bundling generative AI models and connecting them to customer records, company data, solution databases, product manuals, and more. This dynamic enterprise search, coupled with generative AI, can outperform almost any human customer-service agent.

But NASA might be going even further.

Its innovative approach of assigning employee IDs to AI systems symbolizes the anthropomorphization of AI, reflecting the readiness to embrace it as an integral part of the team. The Artificial Intelligence Center of Excellence at NASA is at the forefront of blending human expertise and AI capabilities for enhanced problem-solving and efficiency.

NASA's Jet Propulsion Laboratory (JPL) has been using AI to assist in the exploration of Mars. The Curiosity rover, which has been exploring the Martian surface since 2012, uses an AI system called Autonomous Exploration for Gathering Increased Science (AEGIS) to analyze images and autonomously select targets for further investigation. This human–AI collaboration has enabled the rover to make more efficient use of its time and resources, maximizing the scientific output of the mission.

The concern over machines usurping human roles is gradually being overshadowed by the evolving dynamics of human–AI collaboration, emphasizing the unique contributions each brings to the table. Centaur chess, where human strategic insight combines with AI's computational prowess to outperform both human and AI competitors operating independently, serves as a positive example of the vast potential that collaborative systems hold to surmount intricate challenges.

In medical imaging, AI's precision in diagnosing diseases by parsing through millions of images to unearth patterns imperceptible to the naked eye bolsters doctors' diagnostic accuracy. This symbiotic relationship enables medical personnel to allocate more time to patient care and treatment strategy while leveraging AI's analytical might to elevate health-care outcomes.

A study published in the journal *Nature Medicine* in 2024 examined the potential of human–AI collaboration in medical diagnosis, specifically the effects of AI assistance on radiologists. The researchers studied setups in which AI systems worked side by side with clinicians to make decisions regarding medical-image interpretation. What the researchers found was that although AI shows great potential for increasing efficiency, it was important, in effect, for the human and the AI to "understand" each other. The researchers explained that "to optimize the implementation of AI in clinical practice, it is crucial to have a comprehensive understanding of the heterogeneity—the diverse and individualized effects—of AI assistance on clinicians. Clinicians possess varying levels of expertise, experience and decision-making styles, and ensuring that AI support accommodates this heterogeneity is essential for targeted implementation and maximizing the positive impact on patient care."

In other words, just as good human teams need to understand their individual members to perform at their best, the partnership between humans and AI will have similar dynamics. The more these two coevolving entities understand each other, the more completely they will merge. However, this collaboration doesn't stop at individual relationships; it scales into emergent systems capable of previously unimaginable problem-solving. The implications of emergent properties in systems theory stretch into every domain of human endeavor, from innovation to governance, and suggest that the future of development lies in harnessing the power of collective systems. But how would this come to pass? What does it mean for a zettabyte to be "second nature" to a supermind?

Imagine a future in which vast distributed networks of AI, linked to humans through brain–computer interfaces, work together to solve complex problems. One real-world example is NASA's Jet Propulsion Laboratory, which is already integrating AI with human experts to tackle deep-space mission planning. Here, the AI can simulate millions of potential mission paths based on enormous datasets of planetary information, allowing humans to focus on the most creative and high-level decisions. Similarly, Deloitte has employed "cognitive technologies" to aid in decision-making, sifting through immense datasets at speeds unimaginable to a human. This kind of collaborative intelligence shows how humans and machines can come together to solve problems at scales and speeds beyond human capacity alone.

Now consider how such systems, like superminds, could handle the global datasphere. By 2025, the datasphere is expected to reach 175 zettabytes. To put this into context: all the words ever spoken by humans are estimated to total 42 zettabytes. Yet for a supermind—an integrated system of human–AI collaboration—processing zettabytes of data is not a challenge but a natural capability. Take, for instance, the complex algorithms used by Google to organize, filter, and serve search results from the vast expanses of internet data, or the efforts of CERN's Large Hadron Collider to sift through petabytes of particle-collision data to discover fundamental aspects of our universe. Scaling this to zettabytes is simply a matter of system integration and computational power, which superminds would process without hesitation.

These superminds wouldn't just reside in centralized institutions like NASA or CERN; they would be distributed across societies. Cities would use AI-powered infrastructure—think of smart cities with integrated-sensor networks capable of anticipating traffic flows, managing energy use, and ensuring safety in real time. Every decision and action would be informed by zettabytes of data processed collaboratively between human and AI agents. Imagine if the city of Singapore, which already uses AI for traffic management, expanded its network to integrate real-time data from health, environmental, and economic sources to optimize public services on an unprecedented scale.

And what of the emergent properties when humans and AI collaborate at such a large scale that they become a single cybernetic entity? We're already seeing early versions of these systems take shape in areas like customer service, where AI manages repetitive tasks while humans handle creative problem-solving. But imagine this on a more advanced scale, where the collective intelligence of a distributed system—combining the strengths of AI's computational power and human intuition—tackles problems like climate change or global pandemics. AI-enhanced humans, connected through BCIs, could process vast datasets and develop unique insights in real time, directing resources more efficiently than any single government or organization could manage today.

In this emergent future, how we approach collaboration between human minds and AI systems will be critical. The potential of superminds lies in their ability to combine human creativity and AI's data-crunching prowess, creating solutions on scales that transcend what either could achieve alone. We won't just be using AI to automate; we'll also be creating collective intelligences where humans are integrated into a system so profoundly that the distinction between human and machine intelligence blurs.

As automation continues to advance, the uniquely human skills—empathy, creativity, adaptability—will become increasingly valuable. Already, these qualities are being prioritized in job markets where AI performs much of the routine work. However, the true potential of human–AI collaboration lies in collective systems, where human intuition, emotional intelligence, and decision-making are amplified by AI's ability to analyze massive datasets in real time and where both synthetic and biological intelligence merge and coevolve.

Early forms of these superminds are already taking shape. From hierarchical organizations that use AI to streamline logistics (like Amazon's warehouses) to democratic platforms that aggregate collective intelligence (like Wikipedia), these systems demonstrate how human–machine collaboration is evolving. But what if this concept were pushed further—beyond logistics, beyond data aggregation—to include entire social, political, and economic systems? Imagine AI-powered governance where real-time data on public sentiment, resource availability, and environmental conditions inform government policy in a way that adapts as quickly as the problems arise. This is the promise of distributed superminds.

In this future, the collective wisdom of humans and machines will reshape our world in ways we are only beginning to imagine. The whole, as it has always been in systems theory, will transcend the sum of its parts.

CHAPTER 6

CYBERNETIC CONFLICT: HYPERWAR

In the mid-2010s, the notion that warfare could become significantly automated, or even autonomous, was met with considerable skepticism. Many military leaders and defense analysts believed that the complex, dynamic, and high-stakes nature of combat operations would always necessitate human judgment and decision-making. The idea of machines engaging in warfare without direct human control seemed like science fiction at best and a dangerous, unethical proposition at worst. There was much said about humans "in" and "on" the loop.

Even then, I argued that the expectation that there would always be a human in the loop, or even "on" the loop, was unrealistic. I spoke about this at numerous conferences, wrote articles in journals, and even brought this topic up in my first book, *The Sentient Machine*, which was published in 2017. My view has always been that autonomy in warfare is subject to the constraints and demands of game theory. It is an invisible capability that cannot be inspected by satellites or spy aircraft. Therefore, opponents will always assume that their competitors, no matter what they say publicly, are in pursuit of or in possession of full autonomy. If this were the case, then the opponent would have a faster reaction time and would therefore have an immense advantage. I argued that no military would ever allow such an advantage to develop while it "slept at the wheel." Therefore, I saw much of the commentary around keeping humans in the loop merely as convenient posturing. It was never really backed by real intent. It couldn't be.

In fact, the seeds of a more automated and AI-driven approach to warfare were being sown. The US Department of Defense (DoD), through its advanced research arm DARPA, had been investing in AI and autonomous systems for decades. As early as the 1960s, DARPA funded the first academic AI research hubs at MIT, Stanford, and Carnegie Mellon. The Cold War–era SAGE air-defense system, which could process radar data in real time to guide interceptor aircraft, hinted at the potential of automated decision-making in combat.

But it wasn't until the 2010s that the confluence of big data, advanced machine-learning algorithms, and exponential increases in computing power began to make the prospect of autonomous warfare seem more plausible. In 2014 the Pentagon's "Third Offset Strategy" explicitly called for leveraging AI and autonomy to maintain the US military's technological edge. DARPA launched a spate of programs aimed at developing more adaptive, resilient, and autonomous systems for military applications.

Still, many remained unconvinced. A 2015 paper in the *Case Western Reserve Journal of International Law* titled "The Debate over Autonomous Weapons Systems" dismissed autonomous weapons as overhyped, arguing that they were not "artificial intelligence. There will not be 'human qualities' such as . . . semantic understanding. . . . [A]utonomous robots being discussed for military applications are closer in operation to your washing machine." The authors contended that although automation might assist human warfighters, the complexities of warfare would prevent machines from ever fully replacing humans on the battlefield.

Similar doubts were echoed by military leaders. In 2016 then-secretary of defense Ash Carter, though acknowledging the potential of AI, cautioned that "there will never be true 'autonomy'" in warfare. Never, I thought, was a long time. Carter stressed the importance of keeping humans in the loop when it came to lethal decision-making. This sentiment was codified in the DoD's 2012 directive on autonomous weapons, which mandated "appropriate levels of human judgment over the use of force."

Beneath this skepticism, however, the groundwork for more autonomous warfare was being laid. But the successful use of semiautonomous drones for surveillance and targeted strikes in the War on Terror had begun to normalize the idea of machines playing a more active role in combat. Research into swarming tactics, where large numbers of simple autonomous agents collaborate to overwhelm adversaries, hinted at new paradigms for AI-driven warfare.

Strategic competitors were racing ahead with their own autonomous-weapons programs. Russia and China were aggressively developing and deploying AI for military purposes. This raised fears of an "AI arms race" and the specter of autonomous warfare becoming an inevitability, regardless of any qualms.

As General John R. Allen and I pointed out in our October 2021 INSS/PRISM Speaker Series talk hosted by the National Defense University, the potential for fusing the technical "character of war" with the human "nature of war" is increasing as autonomous systems proliferate on the battlefield. We argued that the accelerating pace of technological change is driving us toward a "hyperwar" environment, where decision cycles would be compressed and humans would risk being left out of the loop.

Our work reflected a broader shift in thinking about autonomous warfare that began to take hold in the late 2010s. The question was no longer if autonomous weapons would transform the battlefield, but when and how. The cynicism of the mid-2010s, while understandable, failed to anticipate the speed at which AI capabilities would advance and the pressures that would drive their adoption for military uses. This, in some sense, has been the history of AI development. Humans thought AI would never play chess, much less beat them. But it did. They thought Go was far too complex for a machine mind to fathom. But it did. I was told by vibration analysts and mechanical engineers that the AI that I was building would never offer better insights into machine failure than human experts. But it did. Today, skepticism is focused on whether LLMs can reason or whether autonomous cars can actually be autonomous. And yet again, they will.

As we will explore in the following sections, the realities of autonomous warfare have already begun to manifest in surprising ways, from the skies over Ukraine to the streets of Gaza. The fully automated battlefields once entirely dismissed as mere science fiction have not yet materialized, but we are witnessing the emergence of a new era of warfare, one in which the boundaries between human and machine, biological cognition and artificial intelligence, are increasingly blurred. The implications of this cybernetic fusion will be profound, reshaping not just how wars are fought but also the very fabric of military organizations and the societies they defend.

DOD MEMO

As the 2010s drew to a close, the DoD began to grapple more seriously with the implications of autonomous and AI-driven warfare. This shift was reflected in a series of strategic documents and initiatives that sought to articulate the DoD's vision for the future of warfare and the role of emerging technologies in achieving military objectives.

One of the most significant of these was the 2018 DoD Artificial Intelligence Strategy, which outlined the department's plan to harness AI for a range of military applications, from enhancing situational awareness and decision-making to enabling more-autonomous systems. The strategy emphasized the need to develop "resilient, robust, reliable, and secure" AI systems that could operate in "contested environments" and adapt to changing conditions on the battlefield.

Central to this vision was the concept of "human–machine collaboration," which recognized that although AI could augment and extend human capabilities, it should not replace human judgment entirely. The strategy stressed the importance of designing AI systems that could work seamlessly with human operators, leveraging the strengths of each to achieve optimal outcomes.

This principle was further elaborated in a 2020 DoD memo titled "Ethical Principles for Artificial Intelligence." The memo laid out five key principles to guide the development and use of AI in the military context: responsibility, equitability, traceability, reliability, and governability. It emphasized that humans must remain responsible for the development, deployment, and use of AI systems and that such systems should be subject to rigorous testing and oversight.

Although the DoD voiced these principles, it continued to push the boundaries of what was possible with autonomous and AI-driven systems. In 2019, DARPA announced the Offensive Swarm-Enabled Tactics (OFFSET) program, which aimed to develop swarms of up to 250 collaborative autonomous aircraft capable of operating in urban environments. The goal was to create swarms that could autonomously navigate, identify targets, and coordinate attacks with minimal human intervention.

Other DARPA programs, like the Squad X Experimentation program, focused on integrating autonomous systems and AI-driven decision-support tools at the tactical edge, empowering small units with enhanced situational awareness and adaptability. These efforts reflected a growing recognition that in the fast-paced, information-saturated battlefields of the future, human cognitive capacities alone might not be sufficient to maintain a competitive edge.

These efforts were echoed by the rapid advances being made by strategic competitors such as China and Russia. In 2017 China announced its "New Generation Artificial Intelligence Development Plan," a comprehensive strategy to make the country the world leader in AI by 2030. The plan identified military applications as a key priority, with a focus on developing autonomous weapons, intelligent command-and-control systems, and AI-driven logistics and support.

Russia, meanwhile, was actively fielding semiautonomous and autonomous systems in real-world conflicts, providing a glimpse of the future of warfare. In Syria, Russia deployed the Uran-9 unmanned ground vehicle, capable of autonomous navigation and equipped with anti-tank missiles and a 30mm cannon. Although the system's performance was mixed, it demonstrated Russia's willingness to test and refine autonomous weapons in live combat situations.

These developments added urgency to the DoD's efforts to harness AI and autonomy for military advantage. In 2018 the department established the Joint Artificial Intelligence Center (JAIC) to accelerate the adoption of AI across the armed services. The JAIC's mission was to "transform the DoD through artificial intelligence," serving as a focal point for AI strategy, policy, and coordination.

As the DoD raced to keep pace with the evolving landscape of autonomous warfare, ethical and operational challenges loomed large. How could the principles of responsibility and human control be maintained in an environment where machines were making more and more decisions? How would the integration of autonomous systems change the nature of command and control, and the role of the human warfighter?

As Deputy Secretary of Defense Kathleen Hicks noted in a 2023 speech on "The State of AI in the Department of Defense," the DoD remained committed to developing AI systems that were "safe, secure, and trustworthy." She emphasized that "there is always a human responsible for the use of force," even as the department worked to integrate AI more deeply into its operations.

As we will illustrate, the realities of autonomous warfare are already testing these principles in profound ways. From the streets of Ukraine to the skies over Gaza, the boundaries between human and machine are blurring, creating new challenges and opportunities for the conduct of war.

But isn't a cybernetic future one in which humans will always be in the loop? How is this push toward human control over autonomous systems antithetical to the core premise of this book? In fact, it is a matter of scale and scope. How large is your cybernetic system? When you think of a cybernetic construct, do you think of a human and a gun modeled as one system? Do you think of a hundred humans, vehicles, and artillery modeled as a single system? Or do you think of five thousand tactical robots controlled by a single company—about a hundred soldiers—as a cybernetic system? It matters. Because in that last case, human cognition and control are only a small part of the decision-making involved in running a cybernetic system of such scale. What I can see happening in the years ahead is that the cybernetic systems being planned in departments of defense the world over, and being built in labs of leading weapons manufacturers, will move the slider toward a larger number of machine-originated actions and a smaller number of human-originated actions. From our cybernetic lens, this is precisely why we want to shift the debate from what machines can do and what humans can do to what cybernetic systems can do as a whole. Whether we like it or not, we are indeed at the cusp of a new era of cybernetic warfare. And the international community's efforts to navigate this uncharted terrain will be critical in shaping the future of armed conflict.

THE PREDICTIONS COME TRUE

As the 2020s dawned, the prospect of autonomous warfare was no longer a distant hypothetical but an emerging reality. This shift was driven not only by the rapid advancement of AI and autonomous technologies but also by the changing nature of conflict itself. Nowhere was this more apparent than in the ongoing war in Ukraine, which had become a testing ground for a new generation of autonomous weapons and tactics.

In an article I wrote for *Forbes* in late 2021, I outlined a series of predictions about the role of AI and autonomy in the Ukrainian conflict. I argued that the war in Ukraine would accelerate the development and deployment of autonomous systems as both sides sought to gain a tactical edge in an increasingly complex and contested battlefield.

One of the key areas where I expected to see significant innovation was in the use of drones and unmanned aerial vehicles (UAVs).

I predicted that the conflict would see the emergence of "air-to-air engagements" between drones as both sides sought to establish dominance in the aerial domain. This would represent a significant shift from the traditional use of drones for surveillance and for targeted strikes against ground targets.

I also foresaw the rise of "drones deploying from other drones," with larger UAVs serving as mother ships for swarms of smaller, more specialized drones. These swarms could be used for a variety of missions, from reconnaissance and targeting to electronic warfare and kinetic attacks. The ability to deploy and coordinate swarms of drones from the air would give commanders new options for projecting power and engaging adversaries in complex, multi-domain operations.

Another area where I anticipated significant developments was the integration of unmanned ground vehicles (UGVs) and UAVs. I held a conviction that the Ukraine conflict would see the first real instance of "UGV–UAV combined operations," with ground and aerial robots working together to conduct reconnaissance, identify targets, and deliver precision strikes. This kind of cross-domain autonomy would allow forces to operate more efficiently and effectively in urban and other complex environments.

Perhaps most concerning to me was the potential use of autonomous swarms in the Ukrainian conflict. I wrote that "we are likely to see huge numbers of such autonomous weapons all simultaneously looking for targets," creating a new level of lethality and unpredictability on the battlefield. The prospect of swarms of AI-driven machines engaging in combat without direct human control raised profound ethical and operational questions.

As it turned out, many of my predictions would turn out to be correct. In the years following the article's publication, the war in Ukraine did indeed become a proving ground for a new generation of autonomous weapons and tactics. Both Russian and

Ukrainian forces deployed drones and UAVs in unprecedented numbers, using them for everything from reconnaissance and targeting to electronic warfare and kinetic attacks. Footage of Russian drones taking on Ukrainian "Baba Yaga" drone bombers is so commonplace that it can easily be found on X and YouTube. In terms of scale, the online publication *Inside Unmanned Systems* quoted a Ukrainian official claiming that Russia was using forty thousand drones per month just in the FPV (first-person-view) category. Ukrainian forces also made extensive use of tens of thousands of commercial quadcopters, particularly the Chinese-made DJI Mavic 3. These nimble drones proved invaluable for spotting Russian positions and directing precise artillery fire. On the Russian side, the indigenous Orlan-10 UAVs served as the eyes in the sky, providing real-time intelligence and acting as communication relays for other drones, such as the Lancet, in their arsenal.

Electronic warfare has emerged as a crucial aspect of drone operations. Russian forces continue to use sophisticated jamming systems that have wreaked havoc on Ukrainian UAVs. By the summer of 2022, a report from the UK's Royal United Services Institute (RUSI) suggested that nearly 90 percent of Ukraine's drones were being neutralized, their GPS navigation confused and radio links severed by Russian electronic countermeasures.

The kinetic capabilities of drones were fully realized as both sides adapted them for offensive operations. Ukrainian forces ingeniously modified commercial drones to drop mortar rounds on Russian troops and vehicles, and Russian forces countered with short- and long-range loitering munitions, including precision-guided drones designed to strike Ukrainian armor and artillery.

One of the most striking developments was the use of naval unmanned surface vessels (USVs) to attack the Russian Black Sea fleet. Another was the mass use of Iranian Shahed drones by the Russians to target Ukrainian installations in large numbers.

These systems were often deployed in an autonomous mode that prevented Ukrainian electronic warfare from impeding their missions. The Turkish-made Bayraktar TB2 drone was also employed heavily by Ukrainian forces in the early months of the conflict. The TB2, which can be armed with laser-guided bombs and anti-tank missiles, proved to be an effective system while Ukraine could still provide some cover for it, allowing Ukrainian forces to strike Russian targets with precision and relative impunity. The drone's success in Ukraine sparked a global interest in the platform, with several countries expressing interest in acquiring their own TB2s.

Meanwhile, Russian forces deployed a range of autonomous and semiautonomous systems, including the Uran-9 UGV and the Orion UAV. These systems were used for reconnaissance, targeting, and strike missions, often operating in coordination with manned assets. The Orion, in particular, demonstrated the potential for air-to-air engagements between drones, successfully destroying a Ukrainian helicopter drone in a test of its air-to-air capabilities.

The Ukrainian conflict also saw the first large-scale use of drone swarms in combat. In 2022 Russian forces deployed a swarm of KUB-BLA loitering munitions against Ukrainian positions, demonstrating the potential of autonomous swarms to overwhelm defenses and deliver precision strikes. Although the effectiveness of the swarm was limited, it provided a glimpse of the future of autonomous warfare.

As the conflict progressed, the integration of UGVs and UAVs became more sophisticated, with both sides using ground and aerial robots to conduct combined operations. Russian forces used the Uran-9 and Orion in tandem to identify and engage Ukrainian targets, and Ukrainian forces used the TB2 and other drones to provide real-time intelligence to ground units.

The war has also accelerated the development and deployment of autonomous systems, with small factories and workshops springing up all over Russia and Ukraine to build drones of a variety of types. This is a precursor to decentralized military manufacturing, which I am personally tracking as an important trend that will become a subject of future military innovation. The idea that automated systems and Industry 4.0 technologies—such as computer-controlled mills, 3D printers, and general-purpose microcontrollers—can be used to build various weapon systems in small, distributed environments offers a significant logistical advantage. Industry 4.0, a reference to the fourth industrial revolution, enables smart, flexible manufacturing through advanced, interconnected technologies, making it feasible to produce complex items like weapon systems efficiently and in decentralized locations. This capability allows countries and groups that cannot fully protect against aerial attacks to maintain their weapon supply, even if one manufacturing site is compromised. Militaries are trying to adapt to this new reality, but true to form for any large enterprise, they are doing it slowly. In order to be prepared for the next conflict, militaries do have the professional imperative to learn from Ukraine and begin developing new doctrines, tactics, and technologies to harness the power of autonomy while mitigating its risks.

My friend Pravin Sawhney, a retired Indian Army officer and now the editor of India's premier defense magazine, *FORCE*, is one of the analysts covering the nature of these changes. His advice to the Indian armed forces after observing the Ukraine conflict was "create three new domains of cyber, electromagnetic spectrum & space. . . . In war what will matter is having hypersonic weapons & long range anti ship missiles meshed with surface & undersea drones."

As of September 2024, the tragic conflict in Ukraine continues to grind on. At the time of this writing, the role of autonomous systems and AI-driven technologies has become increasingly central to the conduct of the war. Both Russian and Ukrainian forces have now spent two years perfecting the development of tactics, techniques, and procedures, and they are producing thousands of drones per month. A transformation of the battlefield in this way would have been unimaginable just a decade ago.

As we will explore in the next section, the lessons of Ukraine are already being applied in other contexts, from conflict in the Middle East to Great Powers competition in the Asia-Pacific. As the boundaries between human and machine continue to blur, the implications for the future of cybernetic warfare are far-reaching.

AUTONOMOUS TARGETING IN GAZA

The Gaza Strip, a small, densely populated enclave on the eastern coast of the Mediterranean Sea, has long been a flash point of conflict between Israel and Palestine. In recent years this conflict has taken on a new dimension with the introduction of Israel's autonomous-weapon systems and AI-driven targeting technologies, which have made warfare even more deadly.

Israel has developed and deployed autonomous weapons in Gaza since 2021, continuing its long history of using the area to test new military innovations. Euronews and the AP have both reported that Israel deployed "AI-powered robot guns that can track targets" at various checkpoints, including those in the West Bank. The Israel Defense Forces (IDF) has also used a range of UAVs, UGVs, and other autonomous platforms to conduct surveillance, reconnaissance, and strikes against Palestinians.

One of the most significant developments has been Israel's use of the "Harop" loitering munition, a drone-like weapon that can circle over a target area for hours before launching a kamikaze-style attack. The Harop, which is equipped with a high-explosive warhead and an electro-optical sensor, has been used to attack targets in Gaza as well as to destroy infrastructure. By April 2024, Israel had already dropped more than 70,000 tons of bombs on Gaza and, running low on bombs, had been rearmed with more than $24 billion in aid from the United States. Part of the rearming involved a $150M order placed with IAI, the manufacturer of the Harop, to renew the fast-depleting supply of this loitering munition.

Israel has also deployed a range of smaller UAVs, such as the "SkyStriker" and the "Orbiter 1K," which can be used for reconnaissance and targeted strikes. These drones, which are often operated in swarms, have been used to identify and attack targets in densely populated urban environments, where traditional military operations are often hampered.

In addition to aerial vehicles, Israel has used UGVs in Gaza, such as the "Guardium" and the "Jaguar." These remote-controlled vehicles, which are equipped with cameras, sensors, and weapons, have been used for border patrol and kinetic operations. The Guardium, in particular, has been used to police the Gaza border and enforce blockades.

Perhaps most controversially, Israel has also used AI-driven technologies to identify and track potential threats in Gaza. The IDF has developed a sophisticated network of sensors, cameras, and other surveillance systems that can monitor the movements of individuals and vehicles in real time. These data are then analyzed by AI algorithms that can identify patterns of behavior and flag potential threats for further investigation.

Amid the high number of civilian casualties and global calls for ceasefire, human-rights groups such as Human Rights Watch (HRW), Access Now, and Euro-Med Human Rights Monitor have all raised concerns about the use of these technologies, arguing that they have been used indiscriminately and violate the human rights and civil liberties of Palestinians in Gaza. There are also concerns about the potential for these early AI systems to per-petuate biases and make mistakes, particularly when it comes to identifying and targeting individuals.

Despite these concerns, unless internationally recognized laws of armed conflict—specifically, proportionality and the protec-tion of civilians—are enforced, the use of autonomous systems in Gaza is likely to continue and expand in the coming years. The IDF has invested heavily in these technologies, seeing them as a way to reduce the risk to Israeli soldiers and civilians when in conflict with Palestinians.

Meanwhile, in Gaza, Hamas has also sought to develop its own cybernetic capabilities, albeit on a much smaller scale. It has used small drones for reconnaissance and propaganda purposes, as well as to drop incendiary devices across the border into Israel. Although these drones are relatively unsophisticated compared to Israel's, they demonstrate the potential for even nonstate actors to use cybernetic technologies to level the playing field. Today, many of these drones are like a human being extending his or her vision and reach via a flying appendage. The cameras in the drone relay back to a set of goggles a series of views that make it seem that it is the human who is flying about. When an obstruction appears sud-denly, the human operator, miles away from the scene, might duck or weave as if the obstruction were physically in front of him or her. This is a proto-cybernetic extension of a human: senses and some ability to cause action at range.

If Palestinian and Israeli drones are seen as an early form of cybernetic weapon system in this way, it is safe to say that the Palestinian systems have a mostly human component, whereas the Israeli systems have a much greater machine-autonomy com-ponent. Over time, this sliding scale will move more and more toward the machine element doing most of the thinking and the acting. The question for many covering this field has been whether it should. The question for me has always been what we should do when this inevitably comes to pass.

The use of autonomous systems in Gaza raises questions about the conduct of warfare in the twenty-first century. As AI-driven technologies become more sophisticated and widespread, we are building cybernetic militaries. And these cybernetic systems have the human–machine slider tilting more and more to the machine side of things. At great scale, my feeling is this is what will happen with most cybernetic systems. There is just too much intelligence and capability that can be manufactured synthetically, and there is a strong incentive to protect the human who is integrating himself or herself with the weapon systems. Consequently, there is a risk that the use of cybernetic weapon systems will violate human rights and escalate conflicts in ways that are difficult to predict or control.

The scariest part, perhaps, is that despite the bone-chilling destruction we've seen, the use of autonomous systems in Gaza is but a microcosm of the broader challenges posed by the rise of AI in warfare. As these technologies continue to evolve and proliferate, it will be up to international organizations, national policymakers, military leaders, and civil society to ensure that they are developed and used in a way that is consistent with international law, human rights, and the principles of just war.

LONG-TERM AUTONOMY

As militaries implement the lessons of what they see in Gaza and in Ukraine, a major challenge on the horizon is ensuring that their cybernetic weapon systems can operate effectively over extended durations with minimal human supervision. Current autonomous systems rely heavily on rules, constraints, and human oversight to function safely and as intended. But from a military standpoint, the truly game-changing potential of autonomy lies in its ability to operate independently for long periods across wide areas.

Achieving robust long-term autonomy requires overcoming several key hurdles related to machine learning, planning, replanning, and system resilience. Work done by Peter Stone and associates at UT Austin has explored some of these challenges through their efforts in the RoboCup@Home competition and the Building-Wide Intelligence (BWI) project for deploying service robots.

One major focus has been enabling autonomous systems to construct an accurate "world model" from sensor data to understand their environment and mission context. As described in the group's paper, this requires capabilities like semantically mapping the surroundings to associate precepts with human-recognizable landmarks and objects. Their Pose Registration for Integrated Semantic Mapping (PRISM) system aims to accomplish this by extracting semantic information from objects and signage to automatically annotate the robot's map.

With an accurate world representation, the system can then plan intelligent sequences of actions to accomplish goals within that context. Stone's group developed techniques for representing both the robot's current knowledge and hypothetical information from humans in a semantic network. This allows reasoning over uncertain knowledge to find feasible plans and trigger replanning or error handling if assumptions prove invalid.

Throughout planning and execution, resilience is critical in order to handle unexpected events or failures. The UT Austin researchers employed a multilayered architecture with reactive components like hierarchical finite-state machines to quickly adjust under uncertainty, combined with more deliberative planning and lower-level skills. Redundancy and bio-inspired principles could further enhance robustness.

Although many open challenges remain, work like that done by Stone's group has pioneered approaches for key facets of long-term autonomy. The group's multipronged efforts spanning competitions, deployments, and controlled experiments have yielded architectures and algorithms addressing world modeling, knowledge representation, planning, and resilient execution—all crucial to realizing truly autonomous systems.

NEXT TIME, IT'S GOING TO BE DIFFERENT

When General John R. Allen, Deputy Secretary of Defense Robert Work, and I conceptualized and developed thinking around hyperwar many years ago, we emphasized two critical factors: precision and magnitude.

Historically, the lack of precision necessitated the use of magnitude, such as during World War II, when hundreds of bombers were needed to drop thousands of unguided bombs to destroy a single bridge. Today, precision has evolved to a point where a single JDAM (joint direct attack munition) smart bomb can accomplish the same objective with unparalleled accuracy.

The development of swarm intelligence and the underlying physical systems necessary to deploy swarms in large quantities have reached a level of sophistication that will fundamentally change warfare. Ukraine is a preview, with large numbers of drones being used but most of them under human control. The future will be different: drones of varying sizes will deploy other drones, and all of them will orchestrate action autonomously. In one crucial way, drone swarms are reminiscent of the early days of nuclear testing. Just as nuclear weapons represented the extreme of magnitude with their capacity for indiscriminate destruction, tomorrow's drone swarms epitomize precision at scale and low cost.

Facing these swarms should invoke the same realization that nuclear weapons once did: the principle of mutually assured destruction (MAD). The difference of course is that mass precision can destroy vast infrastructure even with a much smaller amount of explosive, kinetic force. Why blow up a tank through its armor when you can fly down the barrel of its main gun and explode inside? Can the mass deployment of such precise, autonomous systems render conventional warfare impractical, much like the threat of nuclear annihilation did for large-scale conflicts between superpowers? My hope is that AI and autonomous systems will make kinetic conventional warfare as impractical and obsolete as nuclear weapons have made total war.

However, when I discussed these ideas with General Jack Shanahan, the former head of DoD's famous Project Maven and the founding director of the DoD's Joint Artificial Intelligence Center (JAIC), he was skeptical. He argued that war, by its very nature, defies ultimate conclusions and will persist as long as there are people willing to engage in conflict. No matter how advanced technology becomes, the human element of war ensures its continuance in some form. I suspect that he is right and that the precise shape the wars of the future will take will most certainly surprise us all. As far as I can see, war, even with fully autonomous systems at work, will remain a cybernetic construct on the whole. The human element will be in there somewhere, although the human relationship with autonomy will continue to change.

We can see this phenomenon in play with the next generation of aircraft weapon systems. At one time, fighter pilots would talk about being one with their aircraft in a very physical, kinetic sense. Of feeling its maneuvers and anticipating its acceleration. They merged with technology in order to create a cybernetic system in the sky that was optimized for maneuver: to establish a position behind the enemy. But as the maneuvering has been increasingly offloaded to machines, the cybernetic relationship between fighter and pilot has evolved. Advancements in very-long-range (BVR) missiles, like the AIM-260 and AIM-174, with ranges from 250 to 400 kilometers, and their Chinese counterparts, such as the PL-15, PL-17, and PL-21, have reshaped air combat and the expectations of what a pilot must do to excel.

Pilots no longer need to train as intensely for dogfights but instead for a new kind of situational awareness and decision-making in conjunction with AI and drones. Today's pilots, equipped with extensive sensory augmentation and decision-support systems, are already quasi-cybernetic entities, yet they are different from the generation that came before. The human element is shifting to the more strategic elements of combat, whereas the machine autonomy is taking care of the tactical elements more and more. And the future will see even greater integration of autonomous elements. We will go from autonomous missiles to autonomous wingmen, such as the Turkish supersonic Kizilelma unmanned combat aerial vehicle, within fighter formations.

This shift toward autonomous systems doesn't just enhance the potential for parallelized destruction but also amplifies the decision-making capabilities of individual operators. Who would suggest that the symbiotic relationship already emerging between human warfighters and intelligent machines won't transform the dynamics of warfare?

THE FUTURE OF HUMAN-MACHINE SYMBIOSIS IN WARFARE

As artificial intelligence capabilities advance, the integration of intelligent machines as trusted partners alongside humans will become increasingly viable and critical on the battlefield. Rather than full autonomy in all areas, a symbiotic model leveraging the complementary strengths of people and machines may be the optimal path forward in many cases.

Machines possess advantages over humans in areas like raw computing power, massive data-processing and pattern-recognition abilities, and tirelessness. They can digest sensor feeds, communications traffic, intelligence on enemy dispositions, and other data streams at rates incomprehensible to human analysts. AI planning algorithms can rapidly explore the decision space and generate high-quality options across various scenarios. And autonomous robotics can undertake laborious or dangerous tasks without fatigue or risk of casualties.

Conversely, human warfighters bring irreplaceable assets like physical resilience, commonsense reasoning, rich social/emotional intelligence, and skilled improvisation. They can apply hard-earned combat experience and adapt creatively amid the chaos and unavoidable ambiguities of war. Distinctly human qualities like courage, loyalty, and self-sacrifice will remain vital. And they do have something to lose. That sense alone can be a double-edged sword, encouraging both safe action and incentivizing cruelty in the name of self-preservation.

Rather than replacing humans, intelligent machines can instead amplify their abilities by serving as smart tools and cognitive partners. Sensor data could be continuously analyzed, filtered, fused, and presented through adaptive user interfaces to provide unprecedented battlefield awareness. AI planners could rapidly evaluate scenarios and courses of action based on high-level human intents. Autonomous systems could respond much more rapidly than humans to direct threats, with a person ultimately validating any use of lethal force.

This potent combination could dramatically accelerate the OODA (observe, orient, decide, act) loop that governs military operations. The OODA loop is a four-step process used for decision-making in all manner of fields, including litigation, law enforcement, business, and the military. The process looks something like this: An individual or organization will first observe the situation, looking at outside information, unfolding circumstances, and the environment and any changes to it. Next, they will orient themselves to the situation, analyzing any previous experiences, new information, or cultural traditions. The third step is to decide, based on the previous two steps, what action to take. The final step is the action. Feedback should be given at each step and used for the next iteration through the loop.

When an individual or organization working through the loop is able to process the steps quickly, it can offer a strategic advantage over any competition. But looking at the steps, it is apparent that careful consideration must be paid at each stage, or faulty decisions are more likely to be made. In the business conference room, bad decisions could result in the loss of millions of dollars. On the battlefield, the wrong decisions could result in the death of civilians or military personnel.

Each decision made on the battlefield abides by the OODA loop. Commanding officers analyze the situation and make these decisions, using available resources. But with using AI as a tool on the battlefield, everything changes. AI's capability to process vast amounts of data and to execute complex algorithms enables militaries to make more informed and rapid decisions, thereby compressing the OODA loop to near-instantaneous responses. Humans could make well-informed decisions more quickly based on machine intelligence. Unmanned systems could then execute kinetic actions or cyber-actions almost instantaneously. This cybernetized command and control would be critically empowered to outpace and disrupt potential adversaries.

Of course, introducing intelligent-machine partners raises substantial challenges around human–machine communication, trust calibration, training paradigms, verification and validation of machine outputs, and more. User experience and human factors considerations will be crucial to seamless teamwork between people and AI/robotics systems in high-stakes environments.

Ensuring robust machine ethics, alignment with human values, and meaningful human control over key decisions like use of force will be essential. But when we use technologies of this type, we also expand the potential for exploitation. Adversarial attacks on AI components must be studiously guarded against. In corporations, these attacks can bring down communications systems and halt operations. On the battlefield, these attacks cost lives.

Human–machine symbiosis on the battlefield could be a revolution in lethality and strategic advantage. It would be the embodiment of network-centric warfare, with every asset and human unified through ubiquitous networking and advanced data fusion into a highly coordinated war-fighting organism. The side best able to cultivate this cybernetic warfare paradigm may decisively overmatch opponents limited to purely human or autonomous capabilities. Humans may not be enough. AI may not be enough. Only a combination of the two may prevail.

Perhaps on the bright side, the technologies and techniques pioneered to enable symbiotic human–machine combat teams would pay tremendous dividends in other domains. Fields like assistive robotics, augmented reality, self-driving vehicles, and even cognitive prosthetics could all evolve more rapidly given advances toward true machine partnership with people in extreme circumstances.

HUMAN–MACHINE FUSION
THROUGH EXOSKELETONS

Separate from the development of unmanned vehicles of many types, the human warfighter is also undergoing a profound transformation.

One significant development in this regard is the emergence of human–machine fusion technologies, which seek to enhance the physical and cognitive capabilities of soldiers through the use of advanced robotics, AI, and biotechnology.

At the vanguard of this trend is the development of military exoskeletons, which are essentially wearable robots that augment the strength, speed, and endurance of the human body. These systems, which range from simple mechanical supports to fully powered suits with integrated sensors and AI, have the potential to revolutionize the way that soldiers operate on the battlefield.

One of the most ambitious efforts in this area is the US military's Tactical Assault Light Operator Suit (TALOS) program, which aims to develop a fully integrated combat exoskeleton for special operations forces. The TALOS suit, which has been dubbed the "Iron Man suit" by the media, is envisioned as a highly advanced system that would provide soldiers with enhanced protection, strength, and situational awareness.

The suit would be equipped with a range of sensors and communication systems, allowing soldiers to receive real-time intelligence and tactical information while on the battlefield. It would also feature advanced materials and power systems that would enable soldiers to carry heavy loads and operate for extended periods without fatigue.

The TALOS program has faced technical and budgetary challenges, but it represents a broader shift toward the integration of man and machine in modern warfare. Other countries, including Russia and China, are also investing heavily in exoskeleton technology, seeing it as a way to gain a tactical advantage, amplify human physical capacity, and protect soldiers on the battlefield. The Russians in particular have disclosed their work on the Rostec Exoskeleton, which allows soldiers to hold loads of up to 20kg for extended periods of time without tiring, and allows them to carry up to 60kg while reducing musculoskeletal strain by half.

Exoskeletons are about as cyber-physical as one can get, but even beyond their abilities, there are efforts to develop seamless interfaces between humans and machines, such as brain–computer interfaces that would allow soldiers to control weapons and other systems with their thoughts. These technologies, which we surveyed earlier, are still in the early stages of development but have the potential to blur the lines between human and machine even further.

Exoskeletons and brain–computer interfaces together, with a human at their core, are the ultimate cybernetic combination. And one that isn't very far from being realized.

The fusion of human and machine in warfare raises a host of philosophical questions about the nature of human agency and the role of technology in shaping the future of conflict. On one hand, these technologies have the potential to reduce the risk to human life on the battlefield by allowing soldiers to operate at a greater distance from harm and with enhanced physical capabilities. On the other hand, there are concerns about the potential for these technologies to dehumanize warfare and erode the moral and ethical constraints that have traditionally guided the use of force. There are also questions about the long-term effects of human–machine fusion on the mental and physical health of soldiers, as well as the social and political implications of creating a new class of "super soldiers." In this context, in particular, fans of Marvel will recall the anguish that both Captain America and his sometimes-friend, sometimes-enemy Bucky feels every time the memory of their painful transitions to super soldier enters their minds.

Despite these concerns, my fear is that realpolitik will likely see these concerns being cast to the winds. The development of human–machine fusion technologies for military purposes is likely to continue apace, driven by the evolving nature of warfare and the desire for advantage on the battlefield. It is up to the citizens of every country to require greater transparency and accountability when it comes to the development and deployment of these systems, as well as the establishment of clear legal and ethical frameworks to govern their use. The military, industry, academia, and civil society all have their place in this dialogue in order to ensure that the development of these technologies is guided by a shared commitment to human dignity and the protection of civilians.

As we think back to the wondrous possibilities enabled by a cybernetic city like Neom, we cannot lose track of the disastrous potential of cybernetic wars—particularly if we cannot enforce the laws of armed conflict. It is essential to keep both these possibilities at the forefront of our minds and to work toward a future in which the use of these technologies is guided by a deep respect for human life and the principles of international law. Only then can we hope to ensure that the fusion of human and machine in warfare serves the cause of peace and human flourishing rather than undermining it.

I am sorry to be the one causing you to dive so deep into one of the darker aspects of the fusion of human and machine. Cybernetic evolution will allow us to be at the center of a system that is more capable in all ways. But as warriors, what cause will we choose to be for? For conquest, to inflict suffering, to ethnically cleanse and capture territory? Or to build a future in which we can free ourselves of pain, hunger, and immense levels of disparity? I am sure you hope, as I do, that we will choose the latter.

CHAPTER 7

CLIODYNAMICS
AND CYBERNETICS

When I was growing up, my father used to remind me that "an idle mind is the devil's playground." This may sound like dated advice, but it has served societal stability well over millennia. To put it simply, people need something to do. Technological advancement has moved forward at warp speed over the past few decades. As a consequence, immense opportunities have been given to segments of society that would not have had such opportunities in years past. With the internet, access to information and education has become nearly free. With automation on the farms and in all sorts of physically laborious roles, people now have more free time than ever before. With Zoom and video conferencing, there is freedom for many to work from home. With faster, more integrated transport systems, there are more places to go to and less time to spend dealing with the erstwhile life-threatening logistics of long-distance journeys. To sum it up in a few words, the opportunities and access people enjoy in many developed societies today rival and exceed what the elite in those same societies enjoyed in times past. Technology accelerates the production of well-informed, educated, well-compensated elites who have time on their hands to think about things other than mere survival.

That sounds like a good thing, but societies are a complex system. When you make such radical changes, are you about to experience unintended consequences? What does history tell us about societies that overproduce elites? To understand how the cybernetic acceleration of elite production will affect the future, let's look at the field of cliodynamics.

Cliodynamics is a transdisciplinary area of study that integrates historical macrosociology, cultural and social evolution, economic history, mathematical modeling of long-term social processes, and the construction and analysis of historical databases. It is a portmanteau of "Clio," the Greek muse of history, and "dynamics," which refers to the study of forces that cause changes and movements within systems over time. The term was coined in 2003 by Peter Turchin, a Russian American scientist who is considered the founding father of this field, although its roots can be traced back to the Middle Ages.

Cliodynamics seeks to apply the principles and methods of various disciplines, such as mathematics, computer science, economics, anthropology, and sociology, to identify patterns and regularities in the processes of human history. By analyzing historical data and developing mathematical models, cliodynamics attempts to uncover the underlying mechanisms and dynamics that drive social and political change, including the rise and fall of civilizations, the outbreak of wars, and the dynamics of population growth and decline. Those familiar with the well-known *Foundation* series, started in the 1940s by Isaac Asimov, may feel a similarity to psychohistory, the algorithmic science developed by the mathematician character Hari Seldon that makes general predictions about possible futures for large groups of people. In addition, "The Year of the Jackpot," a short story published in 1952 by science-fiction writer Robert Heinlein, features a similar method of tracking cycles of history and using them to predict the future.

One of the key mathematical tools employed in cliodynamics is nonlinear dynamics, which deals with systems that exhibit complex, often chaotic behavior as opposed to simpler linear systems. Nonlinear dynamics has been successfully applied in fields such as physics, biology, and economics to model phenomena that are characterized by feedback loops, tipping points, and emergent properties. Yes, feedback loops again. Does that remind you of cybernetics? And of reflexivity? By applying these methods to historical data, cliodynamicists aim to identify similar patterns and mechanisms in the evolution of human societies.

The core idea behind cliodynamics is that historical processes are not merely a collection of random events but exhibit patterns and regularities that can be studied using scientific methods. Proponents of cliodynamics argue that by uncovering these patterns and understanding the underlying mechanisms, it may be possible to develop predictive models and simulations that can provide insights into future historical trajectories.

Cliodynamics draws upon a rich body of historical data, including economic indicators, demographic trends, social and political events, and cultural dynamics. By integrating these diverse sources of information into quantitative models and simulations, cliodynamics aims to model the underlying principles that govern the evolution of human societies. This approach challenges traditional historical narratives that often focus on the actions of individuals or specific events, instead seeking to identify broader structural forces and patterns that shape the course of history.

One example of the type of data used in cliodynamics is the Seshat: Global History Databank, which is a comprehensive database of historical and archaeological information spanning several millennia. The Seshat project aims to collect and systematize data on social complexity, political organization, warfare, religion, and other key variables across a wide range of human societies, from ancient civilizations to the present day. By analyzing this data using statistical and computational methods, researchers can test hypotheses about the factors that drive social evolution and identify recurring patterns in the rise and fall of civilizations.

Cliodynamics has gained traction in recent years as advances in data analysis, computational power, and mathematical modeling have enabled researchers to tackle complex historical processes with greater rigor and sophistication. Although the field remains relatively new and continues to evolve, it has already contributed valuable insights into the dynamics of social and political change, and has the potential to inform policymaking, conflict resolution, and the development of more sustainable and resilient societies.

Turchin was born in 1957 in Obninsk, Soviet Union, and initially studied theoretical biology and evolutionary theory before turning his attention to the study of historical processes. Turchin's interest in this field of study was sparked by his observation that historical events seemed to follow certain patterns and cycles, much like the cycles observed in biological and ecological systems. He believed that by applying the principles and methods of complex systems analysis, it might be possible to uncover the underlying dynamics that drive these historical patterns.

Turchin's background in theoretical biology and evolutionary theory has been instrumental in shaping his approach to cliodynamics. He has drawn upon concepts such as natural selection, cooperation, and competition to develop models of social and political dynamics. For example, his work on the evolution of cooperation has explored how factors such as religion, warfare, and social norms can promote or inhibit the emergence of large-scale cooperative societies.

In the late 1990s, Turchin began to develop mathematical models and simulations to study the rise and fall of civilizations, drawing inspiration from disciplines such as ecology, economics, and anthropology. He collaborated with researchers from various fields, including historians, archaeologists, and computer scientists, to gather and analyze historical data.

Turchin's work has focused on identifying and analyzing various factors that contribute to the dynamics of human societies, such as population growth, resource distribution, overproduction of elites, and state formation. He has developed models and theories to explain phenomena such as the occurrence of secular cycles, the rise and fall of empires, and the dynamics of social unrest and conflict.

One of Turchin's key contributions to the field of cliodynamics has been the development of the "demographic-structural theory," which posits that long-term social and political instability is driven by a combination of population growth, resource scarcity, and elite competition. According to this theory, as population growth outpaces economic growth, living standards decline, and social tensions rise. At the same time, an overproduction of elites leads to increased competition for power and resources, further exacerbating social instability. Turchin has used this theory to analyze historical events such as the fall of the Roman Empire and the French Revolution.

One of Turchin's most influential works is the book *War and Peace and War: The Rise and Fall of Empires* (2006), in which he presents a mathematical model that attempts to explain the cyclical patterns observed in the rise and fall of historical empires. He has also published numerous articles and books on topics related to cliodynamics, including *Secular Cycles* (2003) and *Ultrasociety: How 10,000 Years of War Made Humans the Greatest Cooperators on Earth* (2016).

Turchin's work has garnered both praise and criticism from the scientific community. Although some scholars commend his efforts to bring scientific rigor to the study of history, others question the validity and generalizability of his models and the potential for oversimplification of complex historical processes.

Despite the criticisms, Turchin has remained steadfast in his pursuit of developing a unified theory of history that integrates empirical data with mathematical models. His work has inspired a growing community of researchers who are exploring new frontiers in the study of historical dynamics, and his contributions have played a significant role in shaping the emerging field of cliodynamics.

WHAT DOES TURCHIN SAY ABOUT WHAT'S COMING NEXT?

Cliodynamic research has provided insights into potential future trajectories by analyzing historical data and identifying patterns and cycles. Specific predictions are subject to uncertainty and ongoing debate, but some general trends and observations have emerged from the analysis of cliodynamic data.

One of the key findings from cliodynamic research is the existence of secular cycles, which are long-term oscillations in social, political, and economic indicators that span multiple generations. These cycles are characterized by periods of societal integration and disintegration, with alternating phases of stability and instability, prosperity and decline. More recently, Ray Dalio has popularized this idea in his book *Principles for Dealing with the Changing World Order*.

According to Turchin's analysis, many historical societies have experienced cycles of roughly two to three centuries that were driven by a variety of factors, including population growth, resource distribution, elite overproduction, and the strength of central authority.

A key indicator used in cliodynamic research to assess the stability of a society is the "political stress index," which takes into account factors such as income inequality, elite overproduction, and state fiscal health. When the political stress index reaches critical thresholds, it can signal an increased risk of social unrest and political instability. Turchin and his colleagues have used this index to analyze historical data and identify periods of heightened political stress in various societies throughout history.

Based on this cyclical pattern, some cliodynamic researchers have suggested that many modern societies may be approaching a period of increased instability and potential conflict. Turchin himself has warned that the United States and other Western nations may be entering a phase of social and political turmoil, characterized by increasing polarization, economic inequality, and potential civil unrest.

One has to acknowledge that these predictions are based on broad historical trends and are subject to various uncertainties and limitations. The specific timing, duration, and severity of such cycles can vary greatly depending on various factors and interventions, but studying them is instructive nonetheless.

Some critics argue that cliodynamic models oversimplify complex historical processes and may not adequately account for the unique circumstances and contingencies of different societies and time periods. My own point of view on this is that the data are hard to argue with. It may be true that there are underlying complexities, but there is an argument to be made that the data regarding outcomes reflect all these underlying complexities.

In his book *End Times* (2023), Turchin examines the trends of elite overproduction and popular immiseration in the United States, concluding—tragically, for me and my compatriots—that the country is far along the path toward potential violent political rupture. He cites factors such as rising economic inequality, the growing influence of wealth on political power, and the fragmentation of the elite class as indicators that the United States may be headed toward a period of significant social upheaval.

Turchin's analysis of the United States is based on a combination of quantitative data and historical analogies. He points to the growing concentration of wealth and political power among a small elite, the declining living standards and social mobility of the middle and working classes, and the increasing polarization and dysfunction of the political system as signs of a society under severe stress. He argues that these trends mirror those observed in other historical societies on the brink of major upheaval, such as prerevolutionary France or the later years of the Roman Republic.

As with any work that is driven by deep data but is shocking in its conclusions, Turchin's analysis has been met with both praise and criticism. It reminds us that historical patterns and data can provide valuable insights into potential future trajectories, even if specific predictions are inherently uncertain. By combining quantitative analysis with a nuanced understanding of historical context, cliodynamics offers a framework for anticipating and potentially mitigating societal crises.

TECHNOLOGY AND CLIODYNAMICS

The relationship between technology and history is complex. Think about the way that many emerging technologies have the potential to significantly influence the drivers of historical change identified by cliodynamic research. At the very least, technology intersects with cliodynamics in its impact on income inequality and elite overproduction.

Cliodynamic models have highlighted the role of income inequality and elite overproduction as critical factors in driving social instability and political upheaval throughout history. As societies become more unequal and the number of elites vying for power and resources grows, the risk of social unrest and political violence increases.

One way in which technology can exacerbate income inequality is through the phenomenon of "skill-biased technological change," which refers to the tendency of new technologies to favor skilled workers over unskilled workers. As automation and artificial intelligence advance, many routine jobs are being eliminated, but the demand for highly skilled workers in fields such as programming, data analysis, and creative design is growing. This can lead to a widening gap between the wages of skilled and unskilled workers, contributing to overall income inequality.

Although one could point to many sectors of the economy that have played a role, technology and finance are often seen as major drivers of income inequality. The rapid pace of technological change has created enormous wealth for a small number of individuals and companies, particularly in the tech sector. The founders and top executives of tech giants like Amazon, Apple, Facebook, and Google have amassed unprecedented levels of wealth, whereas many workers in other sectors have seen their incomes stagnate or decline.

Similarly, the finance industry has witnessed a massive concentration of wealth in the hands of a relatively small number of individuals and firms. The kind of leverage they enjoy positions them well for big wins. The high salaries, bonuses, and stock options offered by Wall Street firms have attracted a large number of highly educated and skilled workers to the sector, creating a new class of "finance elites" who wield significant economic and political influence.

Another way in which technology can contribute to elite over-production is by lowering the barriers to entry for certain elite professions. For example, the rise of online education and MOOCs (massive open online courses) has made it easier for individuals to acquire the skills and credentials needed to enter fields such as programming, data science, and financial analysis. This democratization of education is generally a positive development, yet it can also lead to an oversupply of elites in certain sectors, intensifying competition for a limited number of top positions.

Tech workers are at least producing products and inventions that are often of tangible value, but the value added by many financial products is often questionable and opaque. This was particularly evident during the 2008 financial crisis, when complex derivatives and financial instruments, often packaged with low-quality or "junk" elements, contributed to the near-collapse of the global financial system.

The growing wealth of both the tech and finance elites, combined with rising levels of inequality and social tensions, could potentially accelerate the process of elite overproduction and contribute to increased instability and political polarization. The sense of entitlement and disconnect from broader society among some members of these elite groups has become increasingly visible, fueling resentment and anger among those who feel left behind.

The irony, of course, is that technology has the potential to be a powerful tool for democratization and empowerment if used and distributed properly. Historically, technologies like the printing press and the internet have played a key role in spreading knowledge and ideas, empowering individuals and communities, and disrupting entrenched power structures. Artificial intelligence can take this to a whole different level, democratizing education, increasing access to markets, reducing product development cost, and so much more.

For example, AI-powered personalized learning systems could help to level the playing field in education, providing high-quality, tailored instruction to students regardless of their socioeconomic background. Similarly, AI and automation could be used to reduce the cost of producing basic goods and services, making them more affordable and accessible to a wider range of people. In the realm of finance, blockchain technology and decentralized finance (DeFi) platforms have the potential to democratize access to financial services and reduce the power of traditional financial intermediaries.

What could governments, educational institutions, and civil-society organizations do to play a more effective role in promoting the democratization of technology and ensuring that its benefits are more evenly distributed? Investments in digital infrastructure and education would help, as well as policies and regulations aimed at promoting competition and limiting the power of tech monopolies. One question to ask is why government spending tends to concentrate in the hands of a few companies: the handful of companies that are the major beneficiaries of all defense contracts and the large tech companies that are vying to concentrate all application hosting for the US federal government on one or two platforms.

There is also a need for greater transparency and accountability in the finance industry to ensure that financial innovations and products serve the wider public interest rather than simply enriching a small elite. One of the trends I am immensely encouraged by is decentralization in finance and the removal of "middlemen" in transactions. These intermediaries often benefit just by being in the middle of a flow as opposed to adding tangible, real value. If security and trust can be delivered algorithmically, then the focus shifts to actual product innovation. Blockchain technology and other decentralized systems could play a role in making this happen.

The democratization of technology and the reform of the finance industry could potentially counteract some of the destabilizing effects of income inequality and elite overproduction by providing more opportunities for social mobility and economic empowerment. If the government begins to facilitate access to decentralized technologies and regulates in favor of such platforms, we may be able to create a more inclusive and sustainable model of economic growth.

Of course, what entrepreneurs and innovators have to contend with is the entrenched power of tech giants and financial institutions. Just look at how much they spend pushing their cause on Capitol Hill! There is most certainly a risk that efforts to democratize technology and reform finance could be co-opted or undermined by vested interests seeking to maintain the status quo. If that is allowed to happen, cliodynamics shows us what is coming next.

By applying mathematical modeling and data-driven analysis to the study of historical patterns and social dynamics, cliodynamics provides a framework for anticipating and potentially mitigating the destabilizing effects of technological change, elite overproduction, and inequality. In a cybernetic system, elite overproduction can be amplified as more and more tech-empowered "superhumans" coexist. And at the same time, perhaps imagined at a citywide or nationwide scale, cybernetic control of human beings shapes their thinking and prevents them from experiencing the disenfranchisement that leads elites to incite turmoil. All that is clear now is that a cybernetic world has the potential to massively amplify the effects predicted by cliodynamics.

The relationship among technology, cliodynamics, and societal outcomes is intricate. Whether technology serves to accelerate or temper the more concerning trends that Turchin and other cliodynamic scholars have predicted will depend largely on the decisions we make—both as individuals and at the institutional level. Our ability to shape how emerging technologies are developed and integrated into society can still determine whether they lead us to more equitable and stable futures or push us toward greater instability.

Of particular concern will be whether we prioritize making technology more democratic and finance more decentralized. If we push for open-source technologies, support community-owned digital infrastructure, and embrace decentralized financial systems that reduce dependency on traditional intermediaries, we may be able to harness the transformative potential of these innovations for the greater good. A future where technology empowers rather than exploits, where access is equalized, and where wealth isn't concentrated in the hands of a small elite is within reach—if we make the right choices.

But there's a flip side. If we allow these systems to further entrench inequality—letting wealth and influence pool in the hands of a technocratic elite—the negative feedback loops that cliodynamics warns about could worsen. Social stratification, political division, and growing discontent could push societies toward instability. We'd be looking at a world where a privileged few reap the benefits of cybernetic enhancements while the rest are left behind, deepening divides. For example, AI and cloud technologies are fast becoming a winner-take-all market. If so, instead of fifty or a hundred people being CEOs of reasonably large AI companies, there is room for only three. In such a world, there is a shrinking set of hierarchies, and they are narrow at the top. There are so many more who are left out than those who are let in. Resistance, misgivings, and mistrust among the neglected elite are guaranteed.

Cybernetic systems could empower individuals in profound new ways or be used for large-scale control and suppression. On one hand, we might see the rise of a "superhuman" class, widening the gap between the enhanced and the unenhanced. On the other hand, there's the potential for these systems to suppress the very tensions that cliodynamics predicts, keeping social unrest at bay— but at what cost?

Navigating this double-edged future will require more than just technological innovation. It means embedding values of inclusivity and fairness into how we shape these new technologies. Public discourse and involvement are key. We need ongoing conversations that include technologists, policymakers, ethicists, and community leaders. Education and digital-literacy efforts will also be critical so that everyone, not just the tech-savvy elite, can have a meaningful say in how our future is shaped. Today, as a technologist, I get the feeling that too many people feel like technology is happening *to* them. Not *for* them. This doesn't bode well for how people will think about technologists: as disconnected promoters of inventions that do nothing but put livelihoods at risk.

By synthesizing the insights of cliodynamics with a meaningful understanding of technological trends, we can work toward building societies that are resilient and adaptable. The road ahead is uncertain, but it is still possible to create a future where technology empowers us all rather than dividing us further.

WHAT CAN WE DO TO ENSURE A BETTER OUTCOME?

Given the potential implications of cliodynamic research and the uncertainties surrounding the impact of emerging technologies, perhaps we should already be considering strategies and approaches that can help ensure better societal outcomes. In complex systems on the scale of global society, there are no guaranteed solutions. Yet several potential avenues can be explored.

One important approach is fostering interdisciplinary collaboration. By bringing together cliodynamic researchers, cybernetic experts, social scientists, and policymakers, we can develop a more comprehensive understanding of the intricate interplay between societal dynamics and technological change. Imagine research centers where experts from diverse fields converge to study the intersection of cliodynamics, technology, and society. These hubs of innovation could become crucibles for new ideas, methodologies, and policy recommendations grounded in a holistic understanding of our complex world.

Adaptive governance and institutional design offer another path forward. By applying insights from cliodynamics and cybernetics, we can create governance systems better equipped to respond to rapidly changing societal conditions. Picture a government with "early warning systems" that monitor key social, economic, and political indicators in real time. Such a system could enable policymakers to address emerging challenges proactively, potentially averting full-blown crises before they materialize.

Promoting societal resilience is equally essential. We must address factors that contribute to instability and vulnerability, such as economic inequality and social polarization. One intriguing concept that has gained traction is "universal basic services." This framework aims to ensure that all members of society have access to essential services like health care, education, and housing. By providing a robust social safety net and investing in human capital, we might mitigate the risks of social unrest and political instability stemming from economic insecurity.

Public education and engagement play a vital role in shaping our collective future. By increasing awareness and understanding of cliodynamic research and technological change, we empower individuals and communities to actively participate in shaping societal trajectories. The Obama administration's "We the People" petitioning system was an interesting attempt at building such a participatory process. Although it was unfortunately discontinued, it demonstrated the potential for digital platforms to facilitate direct citizen engagement with the government. There are many signs in the United States today that the two major parties don't represent the will of large segments of society, particularly the young. The college protests across the nation in early 2024 are ample evidence. To keep moving on as if it is business as usual is to invite the kind of trouble that Turchin has predicted.

Scenario planning and risk-mitigation strategies can help us navigate uncertain futures. The National Intelligence Council's "Global Trends" report is one example of an attempt to anticipate future scenarios. However, its limitations highlight the need for more comprehensive approaches. Perhaps incorporating human–machine collaboration in future iterations could provide a more nuanced and widely informed strategic assessment of global trends.

Ultimately, ensuring better societal outcomes requires us to do many things, not just one or two. We must combine insights from various disciplines to understand the forces at play in our society. We must promote both responsible governance and technological development, and empower individuals to shape their futures actively. We need escape valves and space for people to disagree without being oppressed. In the United States, we cannot be the country that beats and tear-gasses its own students on matters of political disagreement. In Germany and in Europe, the expression of political opinions and speech alone should not be made illegal or invite the use of governmental force. Suppression and a narrowing of the space of discourse don't work for societies in the long term, and the West should know this lesson of history quite well. Yet perhaps Turchin is right, and structural forces are too powerful and inevitable. It remains possible that we will rendezvous at high speed with the approaching train even though we can see it coming our way.

CHAPTER 8

WE ARE ALREADY OPTED IN

By this point, we are all used to being given the choice to opt out of something we might not want online: a newsletter, a marketing list, tracking cookies on a website. If you live in the United States, it's also easy to think that because it's a free country with a Bill of Rights, you are entitled to your privacy and can "opt out" of anything that would violate it.

Let me agree with you on the principle of the thing—before pointing out that enforcing this principle is no longer even remotely viable. All of us are already permanently opted in to the networked reality. And to make a vital point: It will never go back to being like it used to. In fact, it cannot.

Consider a metaphor drawn from evolutionary science. We know that in the age of the dinosaurs—before the Chicxulub asteroid snuffed them out 66 million years ago—Earth's atmosphere altered significantly over time. Besides moisture and temperature changes, ambient levels of oxygen and other gases also went through long periods of rise and fall. The species that were alive at the time adapted to these atmospheric changes, or they died off. (This is one reason that gigantic insects and centipedes no longer roam the land.) The new atmospheric reality didn't have a philosophy, and it was immune to resistance; it simply . . . existed.

That is how this new reality I am describing works. This is the atmosphere we move through. It doesn't matter what you or I think of it—it's still there. The good news is that, unlike the giant centipedes, we need not rely on the haphazard processes of Darwinian evolution for our survival; we can choose how we will adapt to this reality. Perhaps someday a new Chicxulub will wipe us out, too, in the form of a nuclear war, a pandemic worse than we've ever seen, or even another literal asteroid. Meanwhile, though, we can evolve to thrive in this new atmosphere rather than leave our fates in others' hands.

Even if we tried to escape—if we went to live in the woods with no electronic devices—we still wouldn't be able to divorce ourselves from satellites, drones, and all the other technologies that could be used to track us.

Even seeking to minimize use of electronics, to live a more analog life in a digital world, would still leave traces that others could convert into digital exhaust.

In 2013, after the NSA leaks and Edward Snowden's famous defection, FSB, the Russian security service, decided that the only way to ensure security in certain environments, including at the Kremlin, was to revert to using typewriters. I read this news when it came out and was not entirely surprised. But given my technical background, I also wondered how secure typewriters would really be, given the snooping technology we have today. I've been interested in computer security for much of my career, so it is almost a reflex for me that when I read about some new form of protection, my mind begins to consider ways to counter it. When I read about the Kremlin's reversion to typewriters, I thought that they might be more secure than networked computers, but certainly not foolproof.

An incident from my youth sprang to mind. As a boy, I became fascinated with telephones and how they worked. At the time, most phones were based on the pulse dialing system: when you pressed a key or rotated the dial, you would hear audible clicks through the handset. The clicks encoded the number being dialed, with the number of clicks corresponding to the digit. It occurred to me then that someone who wanted to intercept a phone number could eavesdrop from a different extension in the house, listen to the clicks, and decode by ear what the number was. In other words, the dialing of the number created an audio signature that was discernible and predictable. If you learned how to capture and decode it, you could figure out the exact number being dialed.

As I read about the decision to use typewriters in the Kremlin, I wondered: what if you could eavesdrop on the clickety-clack of a typewriter and then decode the audio signature of each character typed? This is where digital processing and machine learning come in. Although every stroke on a typewriter might sound the same to us, they are not truly identical. Each key is mechanically varied, whether by the length of the hammer or even just the shape of the character that hits the plate. Tiny though these variations may be, they are nonetheless variations. Neural networks, the predominant technology used today to build predictive AI models, are particularly effective at processing this type of complex signal, ferreting out and "learning" seemingly imperceptible differences. A neural network that was fed an audio capture of a typewriter would be able to decode the text being typed.

Even then, you might say, wouldn't this require the neural network to train on examples of how each character sounds? Not really, thanks to a curious fact that is common knowledge in the fields of linguistics and cryptography: all languages follow statistical patterns, so there is a frequency with which common sounds and letters are used. The specific frequencies are different for different languages, but the underlying phenomenon applies to virtually every language.

Put these concepts together, and it would be possible to simply capture an audio stream of typewriter sounds, identify the differences between symbols using a neural network, and apply statistical analysis to determine which sound corresponds to which character of the alphabet. You wouldn't even need to know how each key sounded on a specific typewriter to "decode" the audio signature. And you wouldn't even need to have a recording device in the room to capture the sounds. For years now, spy agencies have been able to bounce a laser beam off a window, capture the reflection, and use advanced digital-signal processing to analyze the minor flickers in the reflection—caused by vibrations of the window—to "listen" to a conversation occurring inside the room.

So there's the problem of the Kremlin typewriters cracked. My real point, of course, isn't about typewriters per se but about how modern technology helps us decipher virtually any signal. Analog technology might seem safer for avoiding eavesdropping, but advances in digital signal capture and processing, combined with AI-based signal decoding, can penetrate even such seemingly safe communication methods. Everything we do in the physical world gives off some kind of signature—heat, reflection, movement of air particles—that can be captured and decoded if those who want to listen in have sufficient resources and motivation.

In other words, even if we restrict ourselves to using technology from centuries past, the technology wielded by others inescapably opts us into the new status quo.

Given that even abandoning modern technology for ourselves wouldn't protect us from other parties' data-gathering efforts, just how bad is it when we keep using the connected devices we rely on today? To properly frame the answer, we need to look deeper into the digital exhaust we already emit, new technologies that could make snooping even more invasive, and the routine practices of the technology providers that most of us already use.

The smartphones we carry everywhere already know our location and habits, and in many cases who we're with as well. When you tug down on your iPhone screen and Siri helpfully lists actions you might like to take next, it's because the device is designed to learn your patterns. Say that a malicious actor uses a Strava dataset, which is a collection of GPS data points that is the leading platform for active transportation data, to track your movement, or uses drones or whatever else to predict where you'll be. More patterns. And this doesn't just happen in one-off cases or in thought experiments dreamed up by a computer scientist. It's a routine part of our lives now that we live with such a dense web of connected technology around us.

Maybe you keep your phone turned off when you're away from home to inhibit tracking. That's good—but what about all the traffic cameras and security cameras on buildings that you pass by? Using CCTV footage to track the movements of a victim or a fugitive has become a staple of detective shows and spy thrillers, but there's more than a grain of truth to it. The footage certainly exists; how it's used depends on the motivations of the people or organizations with access to it.

That footage, by the way, is also being augmented by other techniques to make it even more useful for those doing the snooping. For starters, rest assured that the digital-processing and machine-learning tech mentioned in my typewriter thought experiment is already being applied to CCTV feeds. Meanwhile, another idea that seems like science fiction is also ripe to be used in a similar way. In the 2018 blockbuster *Mission: Impossible—Rogue Nation*, Tom Cruise's character and his team of spies have to outsmart many countermeasures, including this new one: gait analysis. Real-world researchers have discovered that they can identify individuals based solely on video footage of how they walk. It turns out that minor differences in our anatomies and biomechanics—bone structure, posture, movement patterns, and so on—create a unique gait "signature" that can distinguish one person from another, just as our retinal patterns or fingerprints might.

Let your mind wander over the possibilities, and it's not hard to imagine a near future where our gaits or other unique movement signatures are layered together and cross-referenced against CCTV footage or datasets like Strava's to identify specific target individuals—with predictable outcomes, including compromised operational security and highly precise kidnappings of VIPs. That's how something that was once just an enjoyable moment in a movie becomes a real-life concern.

There's no doubt that spy agencies, corporations, criminals, and others will get there before we know it. But even without such bleeding-edge technology, snoopers already have a wealth of data to exploit. And much of it comes from the devices and software applications we use constantly. Strava is only the tip of the iceberg, but its nice words about helping users understand their privacy choices are an indicator of the bigger problem: however reasonable it sounds from Strava's or Google's or Facebook's point of view that users already have granular options for privacy and data sharing, it ignores the reality that the average smartphone user isn't very adept at navigating those options.

How many users really take the time to examine and reconfigure the default settings for their phone and every app on it? The entire "Click → Run" workflow is designed to be simple and quick so users will start using an app right away. And the default settings tend to favor the interests of the makers, who want users to share more data, connect the app with everyone in their contact list, and so on. Most users don't interrupt this onboarding process to fully explore the details on some hard-to-reach settings page, let alone check everything all over again anytime they refresh an app or the operating system on the phone.

To get a better idea of the scope of the problem, let's look at some hard figures. According to the analytics firm App Annie, iPhone users are now actively engaging with an average of more than thirty apps per month. Meanwhile, the average number of total installed apps per phone is above eighty. (Keep in mind that in many cases, apps collect data, including location data, even when they aren't being used.) As of 2017, there were already 225 million smartphone users in the United States, or about two-thirds of the entire population of the country. It is simply not reasonable to assume that so many people are going to have the relevant knowledge and take the necessary time to dig into the settings and potential security loopholes for all those apps.

This is why I hold the belief that although much can be done to prepare oneself to live competently in a world full of networked technology, the average citizen will likely not be able to secure and protect themselves entirely. The more sophisticated your understanding, the more prepared you will be. But if history is any guide, not many of us will spend the time to educate ourselves. We may have good intentions, but when it comes right down to it, the majority of us will not follow through.

Unfortunately, even for those who do spend that time in an effort to protect their privacy, it is sometimes not enough. Diligence goes only so far when the technology providers themselves engage in surreptitious data gathering. Exhibit A is Google. An investigation in 2018 by the Associated Press revealed that many Google services for Android and iPhone devices were storing location information even when users had explicitly selected a privacy setting to the contrary. The problem relates to a feature known as Location History; when this is turned on, Google services capture and report the user's location continuously. Google's documentation clearly states that "you can turn off Location History at any time. With Location History off, the places you go are no longer stored." Yet during the course of their investigation, the AP discovered that this was not accurate in practice. Even when users paused the Location History feature, some Google services continued to store data tagged with location coordinates.

How big an issue was this? Well, nearly 2.5 billion Android and iPhone users worldwide were affected by this unwelcome form of data capture. And it took a deep investigation from a major journalistic outlet to expose the problem.

That's why I say that diligence alone isn't sufficient to understand the data exhaust we leave behind, let alone protect ourselves from those who would seek to abuse us. Diligence must be augmented by considerable knowledge. Take the role that Google plays in our lives.

Google's legendary corporate slogan proclaims a commitment to a cause higher than revenue: "Don't be evil." In recent times, however, this slogan has often been lampooned by those who question the real motives behind the company's efforts to track user behavior. At a deeper level, Google's approach makes us wonder whether it is inevitable that the profit-maximizing behavior of the vast corporations at the center of our technological lives will compromise users' privacy interests.

Google's position on this is predictable: the company always claims that it simply wants to understand its users better so it can serve them better. Serve them ads, that is.

Google makes nearly all of its revenue, around 90 percent, by selling increasingly targeted ads, which means that the customers it serves are in fact its advertisers, not its end users. Google therefore needs to show its paying customers how using ads on Google will bring the advertisers more attention and higher sales figures. And the best way Google has found to achieve this is by knowing more and more about every end user's demographics, location, habits, preferences, and decision-making processes—or, if you prefer, about their very minds.

As with Strava, this is a business model that requires a measure of amorality, no matter which high-minded ideals the company claims for itself. Google is essentially an ad network that leverages data about its users to make higher profits. It should not surprise us in the least that it is so intensely focused on learning more and more about users—that is, about us. Every time we do a search, click on a result, watch a YouTube video, or interact with Gmail, the company learns more about us so it can better target us with advertisements.

This logic extends seamlessly to the world of smart devices. A CNET article from October 2018 made the point that as people's habits shift away from traditional search on desktop computers, companies like Google are positioning themselves at the forefront of a new kind of interaction. Instead of typing queries into a search engine, users are now speaking commands to their smart speakers or using their smartphone cameras to identify everyday objects. This allows companies like Google to gather even more personal data—not just about what you're searching for but also about your daily routines, your preferences, and even your environment. The more Google knows about your habits, interests, and preferences, the more effectively it can tailor its advertisements to you, making its platform increasingly valuable to marketers.

Twenty-five years ago, Hotmail was already centralizing information about us by hosting our email online. These efforts, while groundbreaking for the time, look primitive in comparison to the pervasive reach of Google and a few other companies.

Instead of email alone, we now have file storage via Google Drive, Microsoft OneDrive, and other services such as Box and Dropbox. Push your documents to the cloud, and have them available everywhere. For that matter, conduct your meetings online via videoconferencing from Zoom, Google, FaceTime (that's Apple), and so on. All of this is certainly convenient—but it is also quite literally a mechanism by which our most important digital assets and potentially our conversations are being archived on systems we don't control, subject to policies that very few of us understand.

These systems store unimaginably vast pools of data, not only from individuals but also from companies. And at some stage, we can expect these shadows and reflections of our content—stored in data centers we will never see—to come back to haunt at least a few of us. What happens when a hacker gains access to systems like these and walks away with content we didn't even think was online? How about your private correspondence with friends and family members—cordial or bitter, living or dead? What if it's your company's sensitive business plans or intellectual property?

Surely we could delete all of these if we wanted to, right? As someone reasonably familiar with data-center operations, I will tell you that I am not sure that files placed on such cloud-storage mechanisms are even deletable. When you delete things from a cloud drive, the icons representing the files may disappear. You may even tell the system to empty your trash. But this means very little. The automated backup and rollback mechanisms used by each provider vary, but in some cases copies of data we believe were deleted long ago are probably archived on a drive somewhere inside a cloud provider's data center. This won't bother some people. Others may feel an immediate sense of discomfort from knowing they cannot even purge content that is their own.

I've focused on Google because it is so ubiquitous, but the problem is, of course, not unique to Google. Facebook, Amazon, Apple, and other services want to know more about you, too. If you consciously chose to let them, that would be OK—you are an independent node in the network with your own priorities, and how you choose to connect is up to you. But did you actively make that choice? Probably not.

The pervasive surveillance and data collection enabled by the proliferation of connected devices and the centralization of our digital lives are prime examples of how code is being used to shape consciousness and exert control in the modern world. As we generate vast amounts of digital exhaust through our interactions with technology, we are unwittingly, and—perhaps even more importantly—with no viable alternative, providing the raw material for machine-learning algorithms and AI systems to analyze, predict, and manipulate our behavior. You can see this as the erosion of privacy, or you can join the game by using technology to fend off technology. As we'll discuss later, you can allow the increasing power of centralized entities to influence and usurp your thoughts and actions, or you can participate in decentralized alternatives. The extent to which we can be modeled and convinced via digital means raises profound questions about the nature of free will and the power that technology holds over our collective consciousness. And the answers to these questions will become less pleasant if we think that we are opting out of something that can no longer be opted out of.

This may be a good place to emphasize that I have made my career as a serial tech entrepreneur and that I believe strongly in the virtues of commerce. But as an individual—an end user, citizen, husband, father, and friend—I find that I must ask whether the bargains we make, often unknowingly, with these companies are worth it. In exchange for the convenience of universal access, not having to install and run an email server, and getting to avoid basic maintenance tasks such as data backups, we give up our privacy and allow all our communications to be used as data that help advertisers target us. Surely we must ask ourselves: is this a good trade?

It doesn't help that Google is joined atop the Mount Olympus of centralized technology by Facebook, Amazon, Apple, and Microsoft. To Facebook we gladly hand over information on who our friends and family are, where we travel, the activities we take part in, what we care about (from favorite sports teams to hobbies to political groups), and on and on. Keep in mind that the accounting for Facebook includes everything we share on their other platforms, especially Instagram and WhatsApp. Meanwhile, Amazon knows what we read, what we buy for our homes, what we stream via Amazon Prime, and so on. Apple has its hands on iPhone use and location data, music and movies we buy, shows we stream on AppleTV, calls we make via FaceTime, et cetera.

But out of these Olympians, Amazon, Google, and Microsoft hold a special position because their infrastructure hosts a vast array of other companies—including household names like Twitter and business giants like Salesforce—that operate through the now-pervasive software-as-a-service (SaaS) model. This extreme centralization gives these companies their own "cloud gravity," with profound effects on the business landscape.

The Olympians we're talking about, thanks to their enormous scale, are superb operators of network and computing infrastructure. They leverage their scale to reduce the cost of each increment of capacity that they add. They then sell that capacity to other companies, which then use it to deliver SaaS products or simply as an extension of their own operating infrastructures. It doesn't take very long at all for the client companies to become thoroughly enmeshed in the cheap, efficient infrastructure offered by Google, Amazon's AWS service, or Microsoft's Azure service.

Sure, you can still develop software products with a small team, especially when you employ open-source infrastructure to speed up the process. But then comes deployment, which doesn't come cheap. In practice, once it reaches the point of deployment, even the scrappiest software start-up cannot afford to do business any other way than to use these services: replicating the infrastructure would be prohibitively expensive.

Microsoft, Amazon, and Google wield vastly disproportionate power in the tech landscape, even apart from the data they collect on end users, because they effectively form an oligopoly that smaller tech companies cannot afford to eschew. With that in mind, it's not surprising that these three Olympians enjoy huge growth from their cloud-infrastructure businesses, with revenues often climbing by 35–45 percent annually. To give one example that puts it in hard figures, Microsoft's revenues from its Azure operations were $17 billion in just the first quarter of 2021.

Everywhere we look, the few companies that dominate the cloud have embedded themselves into our personal lives, into the companies that serve us, and into our economy as a whole.

The companies we've been talking about, for all their influence, are at least based in the United States. Therefore, it might be some comfort to know that they are subject to US laws—even if they also lobby strenuously to shape those laws in their favor. If Google or one of the others behaved too badly, it would be subject to disciplinary measures: criminal penalties, regulatory oversight, and so on.

But what if I told you that an organization headquartered in a foreign country is, right at this instant, collecting real-time data about the locations and movements of millions of Americans? That it is monitoring our highway network and all the main arteries of our largest cities? And that in fact this organization is collecting information at a finer level of granularity than any of our law enforcement agencies? The information it possesses can even be used to associate individual identities to their places of residence, places of employment, social habits, preferred drugstores and restaurants, and almost every aspect of their daily movements.

That organization is the company Waze, which is based in Israel. Usually, we don't think of Waze as anything other than convenience. The app identifies traffic jams, points out alternate routes to make your morning commute a bit more tolerable, and even projects a sense of social collaboration, given that any Waze user can identify and share traffic accidents, obstructions, detours, and even the location of police patrols and speed traps. All of that information, framed in a user-friendly interface, is genuinely handy anytime you need to deal with the frustrations of traffic. But it's also an enormous trove of data that is managed, owned, and used by a foreign company not entirely subject to US jurisdiction. It is yet another example of the emergent complexity in our networked world—a world in which hugely popular apps like Waze and TikTok have become interwoven with how we live, work, travel, and entertain ourselves yet don't fit within traditional national boundaries.

As with the Strava example mentioned previously, companies like Waze that gather this type of information typically assure users that they don't know much about individual identities because they allow both accounts and the information associated with them to be anonymized. These assurances have the merits of being both convenient and technically true. In practice, however, they are meaningless.

Once you have someone's driving data over time, you can easily identify their place of residence. In fact, mobile assistants such as

Google Now and Apple Siri already use pattern-finding techniques like these to "guess" when you are at home or are ready to head home. Of course, once you know where someone lives—even if the data logs are, technically speaking, anonymized—obtaining their identity from publicly available tax records and the like is trivial. And now we're back to the scenario of a clever housebreaker or a foreign agent being able to collate information from a few disparate sources (not a hard task, these days) to zero in on potential victims. Thus, we once again find ourselves confronted with the reality that amoral companies are constantly collecting data on us to offer us useful services (accurate search results, tracking of workouts, better driving routes), yet in doing so they are either attempting to steer our behavior or increasing the risk that some bad actor can use that data to carry out their harmful plans for us.

But what about the strenuous efforts of lawmakers to address data privacy? Let's consider the case of the General Data Protection Regulation (GDPR), passed by the European Union in 2016. This regulation put into place guidelines for how companies could collect and process personal data and is considered the toughest security and privacy law in the world. GDPR became a huge practical headache in early 2018 as the enforcement deadline loomed into view. Companies doing business in the European Union—meaning virtually all multinational corporations of any size—had to conform to the sternly worded new rules about the housing and handling of customer data. GDPR was a well-intentioned effort to protect consumers' privacy, and at the time companies were rightly concerned about how it would affect their way of doing business. After all, how do you market your wares effectively in the twenty-first century if you can't track enough data about your customers and prospective customers? The business press ran countless articles, many of them quite breathless, predicting significant impacts of GDPR not only on consumer-oriented companies but on industrial and business-to-business firms as well. Naturally enough, some companies also saw a business opportunity here: law firms, IT consultancies, and tech companies with specialties in data handling published white papers and offered new diagnostic tools and services to ensure that organizations would meet GDPR requirements and avoid the hefty fines that were threatened for noncompliance.

With the benefit of hindsight, GDPR was a tempest in a teapot. It hardly impeded the biggest data-gathering companies like Google and Facebook, perhaps even making it more expensive for other firms to compete with them. Today, meeting GDPR requirements has been trivialized for most companies—just one more box to check for their IT and legal teams. Niche vendors have sprung up that can manage GDPR compliance on an outsourced basis; they give you some code to incorporate into your website, and voilà—privacy regulation handled.

When even a law as sweeping as GDPR ends up having minimal impact, it's worth it for individuals to remain keenly skeptical about trusting even well-meaning government entities to watch out for us. By all means, we should continue to push for privacy rights, but it would be foolhardy to assume that legislation or regulation alone will make us invisible. It won't because it can't.

The machines already know everything. And that information is being used every day to shape our responses and reactions in the world of ideas and, indeed, in the physical world, too. The question isn't whether it will be used to shape us and our world. The question is this: will that oppress us, or can it be made to do something different?

CHAPTER 9

THE TECHNOLOGIES OF FREEDOM

In early 2024, Sam Altman, the CEO of OpenAI, said, "In my little group chat with my tech CEO friends, there's this betting pool for the first year that there is a one-person billion-dollar company. Which would have been unimaginable without AI and now will happen." There is the sense in Silicon Valley of "the machines can do it"—with "it" being just about anything!

If one person, amplified with technology—a cybernetic entity—can collect and attract so much capital, we will need to rethink everything from how we value and compensate human labor to how we distribute the benefits of economic growth. Doing so will likely necessitate new models of ownership and governance. It may also require novel forms of social safety nets and wealth distribution to ensure that the benefits of technological progress are shared more equitably across society. What it will certainly require is an active choice: for us to take ownership of the technology, to run toward it. And if we make full use of it rather than offer ourselves up to be used by it, we can master it.

Throughout this book, we have explored the myriad ways in which technology is merging with humanity, giving rise to a new cybernetic ecosystem that is reshaping every aspect of our lives. From the pervasive surveillance apparatus that tracks your every move to the nascent smart cities like Neom that promise a more efficient, sustainable, and responsive urban future, the contours of this new landscape are rapidly coming into focus. Technology is merging with us, with companies and with cities. We are on our way to becoming a cybernetic society.

Amid all of this change, a crucial issue remains: How will we, as individuals, choose to engage with this new techno–human reality? Will we be passive consumers, content to let others shape the terms of our technological merger? Or will we be active participants, seizing the tools of this new era to empower ourselves and forge our own path?

The stakes of this choice could not be higher. As we have seen, the cybernetic future that awaits us can be a horror film or one full of hope and promise. The concept of surveillance capitalism, coined by Harvard professor Shoshana Zuboff, describes a new economic order in which human experience is treated as free raw material for hidden commercial practices of extraction, prediction, and sales. In this system our personal data are not only collected and analyzed but also used to shape our behavior and decisions, often without our knowledge or consent. The implications of this are far-reaching: it threatens to undermine our autonomy, manipulate us, and erode the very foundations of democracy. On one hand, the ever-expanding reach of surveillance capitalism threatens to reduce us to mere data points, our every action and preference mined for profit by corporate giants. The rise of autonomous systems and AI-driven decision-making could lead to a world in which human agency is increasingly subordinated to the dictates of algorithms.

Yet at the same time, the technologies of the cybernetic age offer us an unprecedented opportunity to expand our capabilities, tap into new reserves of knowledge and creativity, and shape our individual and collective destinies like never before. The promise of projects like Neom hints at a future where technology is harnessed not for exploitation but for empowerment—where the fruits of innovation are more equitably shared and where societal flourishing is the ultimate measure of progress.

Realizing this more hopeful vision will require more than just wishful thinking. It will demand active engagement, a willingness to roll up our sleeves and shape the development and deployment of emerging technologies in ways that reflect our values and serve our interests. It will require us to cultivate a new kind of technological citizenship, one that is grounded in digital literacy, ethical reflection, and a commitment to the common good.

This new form of technological citizenship is not just about understanding how to use digital tools but also about being able to critically assess their implications and actively participate in shaping their development and deployment. It requires a shift from being passive consumers of technology to becoming active cocreators, working together to build technologies that serve the public good and promote human flourishing.

Most crucially of all, it will require us to embrace and promote what I call the "technologies of freedom": those tools and platforms that enhance rather than erode our autonomy, that empower us to be more than mere cogs in the machine of surveillance capitalism or algorithmic governance. These are the technologies that will allow us to reclaim control over our data, our attention, and our digital identities, and to participate as full partners in the unfolding cybernetic future.

So what are these technologies of freedom, and how can we begin to harness them in our own lives and communities? Let's explore some of the most promising ones available to us today.

The antidote to Big Telco is community-owned connectivity. Decentralized communications are emerging as vital tools in creating resilient, community-driven connectivity. NYC Mesh, a prominent example, showcases how these networks empower local communities by providing reliable internet access without depending on traditional telecom giants. Founded in 2012, NYC Mesh has grown significantly, expanding its network and technological capabilities. The network operates under a community-driven model, with volunteers installing and maintaining rooftop antennas that form the mesh network. These antennas communicate with one another, creating a robust, decentralized system that can bypass traditional internet service providers (ISPs). One of the key advancements in NYC Mesh is the installation of supernodes—high-capacity nodes connected directly to internet exchange points (IXPs). These supernodes enhance the network's capacity and speed, allowing it to support more users and provide faster internet connections. NYC Mesh currently has multiple supernodes, including those at significant data centers like Sabey Data Center at 375 Pearl Street.

The decentralized nature of mesh networks like NYC Mesh makes them inherently more resilient than traditional centralized networks. If one node in the network fails, the others can continue to communicate and route traffic, ensuring that the network as a whole remains functional. This is particularly important in emergency situations, such as natural disasters or power outages, where traditional communication infrastructure may be compromised.

The impact of NYC Mesh on the local community has been substantial. The project aims to bridge the digital divide in New York City, where more than 1.5 million people lack reliable internet access. During the COVID-19 pandemic, NYC Mesh accelerated its efforts to connect as many households as possible, especially those without other internet options. This initiative provided critical connectivity for remote work, education, and access to vital information during the pandemic. By promoting digital inclusion, NYC Mesh has helped numerous low-income households and underserved areas gain access to the internet. The network's community-driven approach ensures that it remains neutral and does not engage in data monitoring, collection, or blocking, which is a significant step toward maintaining user privacy and autonomy.

Despite its successes, NYC Mesh faces risks, including the need for more volunteers and resources to expand its coverage. Regulatory hurdles and competition with established ISPs also pose significant obstacles. However, the project continues to innovate, exploring new technologies and methods to enhance network efficiency and reliability. Looking forward, NYC Mesh aims to increase its presence across New York City, with a particular focus on connecting more buildings and neighborhoods.

The project is also exploring partnerships with other community networks worldwide to share knowledge and best practices, further strengthening the global movement toward decentralized communications. NYC Mesh exemplifies the potential of decentralized communications to democratize internet access and empower communities. As mesh-network technology continues to evolve, projects like these will play a crucial role in shaping a more inclusive and resilient digital future.

The antidote to Big Tech controlling all computing resources is self-hosting: running your own essential digital services on your own computer. Self-hosted data solutions are gaining traction as individuals and organizations seek greater control over their information. These solutions allow users to store and manage their data on personal servers rather than relying on third-party cloud services. This shift is driven by the need for privacy, security, and autonomy in the digital age. One of the pioneering initiatives in self-hosted data is the Solid project, led by Sir Tim Berners-Lee. My wife, Zaib, and I had the opportunity to sit down with Sir Tim and discuss his vision for Solid in great depth. Our mutual friend, Bob Metcalfe, the coinventor of Ethernet, had organized an intimate gathering at a lovely restaurant in South Austin where, after a delicious meal, Tim gave us a view of what he was building next.

Berners-Lee's Solid aims to decentralize data storage and give users control over their personal information through "pods" (personal online data stores). Users can choose to host their pods on existing providers or run their own servers, ensuring complete ownership of their data. Solid supports various open-source servers, including Community Solid Server and Node Solid Server, and integrates seamlessly with other applications and services through standardized protocols.

The Solid architecture is based on the principles of decentralization, interoperability, and user control. It separates data from applications, allowing users to store their data in a location of their choice and grant access to applications as needed. This approach is fundamentally different from the current centralized model, where data are often siloed within specific applications and platforms, making it difficult for users to control and move their data between services.

One of my favorite self-hosting projects is Nextcloud. It's a platform I have been using for many years now, and it continues to get better. It's another prominent self-hosted data solution that combines user-friendly cloud features with robust security and compliance measures. It provides a comprehensive content-collaboration platform, including real-time document editing, chat, video calls, and file sharing.

Nextcloud's emphasis on security includes end-to-end encryption, two-step verification, and detailed user and role management. It supports deployment of various operating systems and devices, making it accessible and versatile for both individual and enterprise use.

For teams and organizations involved in data science, Datalore, by JetBrains, offers a self-hosted platform tailored for collaborative work. It allows teams to manage and analyze data securely within their infrastructure, providing tools for real-time data processing, machine learning, and visualization. This platform ensures that sensitive data remain within the organization's control, enhancing both security and compliance.

Self-hosting data can minimize the risk of unauthorized access and data breaches, for users are not dependent on third-party cloud providers. Encryption and secure access controls further protect sensitive information. Users retain full ownership of and control over their data, deciding how it is stored, accessed, and shared. This sovereignty prevents data exploitation by corporations for profit. Self-hosted solutions can be tailored to meet specific needs and integrated with existing systems. Users can modify and extend functionalities as required.

Although self-hosting data offers numerous advantages, it also presents challenges such as the need for technical expertise, initial setup costs, and ongoing maintenance. However, many self-hosted platforms provide comprehensive documentation, community support, and user-friendly interfaces to mitigate these challenges. Additionally, advancements in technologies like containerization (e.g., Docker) simplify the deployment and management of self-hosted applications. The shift toward self-hosted data solutions represents a significant step in reclaiming digital autonomy.

Projects like Solid and Nextcloud exemplify the potential for individuals and organizations to manage their data securely and independently. As these technologies continue to evolve, they will play a crucial role in the broader movement toward decentralized, user-controlled digital ecosystems.

Easily licensing your digital content is the antidote to your privacy being breached for the benefit of the world's largest companies. Today, a company like Google can come and scrape your website, your blog, and other content you've produced. Then it can use that content to power responses to searches and make money on each result it produces. To be clear, a "result" is the content you produced. Not content that Google produced. Without your content, there would be no Google. But what do you get paid for it? Nothing.

How do we solve this? In 2022 I cofounded a company called Navigate (nvg8.io), which is building an AI-powered crowdsourced data repository. The Navigate platform allows users to contribute everything from data on their purchases to aerial and street-level imagery captured from drones, dashcam footage, anonymized health data . . . pretty much anything. In return for their contributions, users earn a data dividend in the form of NVG8 tokens, the native utility token of the Navigate ecosystem. The reality is that companies like Google owe their success to their contributing communities, which, all too often, see nothing in return. And this is where our vision for Navigate is quite different. Perhaps over time, Navigate will show itself to be a means of income for the economy of the future. By crowdsourcing high-value datasets and rewarding users for their contributions, the company is truly trying to enable a new class of community-powered applications that ensure that users benefit from sharing data rather than doing it for free.

Navigate alone, of course, is not nearly enough. But attempts like it show that it is possible to innovate new streams of income for the cybernetic economy of the future.

As for the economy, in the social and political realm we will also be compelled to grapple with profound questions about the nature of agency, accountability, and democratic governance.

Data-sovereignty platforms are transforming the way that individuals and organizations manage and monetize their data. These platforms aim to give users control over their data, allowing them to decide how it is stored, shared, and used. This shift toward data sovereignty is driven by growing concerns over privacy, security, and the ethical use of data. Navigate is a key player in the data-sovereignty landscape, providing a platform where users can pool their anonymized data and collectively profit from the pool. This model empowers individuals to benefit from their data rather than allowing corporations to exploit the data for profit. By enabling users to control and monetize their data, Navigate is helping to democratize the data economy and foster a more equitable digital environment.

The shift toward decentralized data management is gaining momentum. This approach allows individuals to store their data in personal data vaults or on trusted decentralized platforms, enhancing privacy and reducing reliance on big data monopolies. Decentralized data management also facilitates secure data sharing and analysis without compromising privacy. Privacy-enhancing computation techniques, such as secure multiparty computation and federated learning, are becoming more prevalent. These methods allow data analysis and AI computations to be performed without exposing raw data. This ensures that data remain private and secure while still enabling collaborative projects.

The regulatory environment around data sovereignty is evolving rapidly. The European Union continues to be the most active organization in terms of pushing regulations such as the General Data Protection Regulation and the upcoming AI Act, both of which may be well intentioned but are also problematic because they are ineffective and discourage innovation by the costs they impose for compliance. I am not certain that these regulations, which impose equally high compliance costs on small and large companies, are the best way to go about it. And it has escaped no one following such regulations that the competence of legislators and regulators in technical matters leaves much to be desired. Still, these regulations influence global standards and push other regions to adopt similar measures. Perhaps someone else can be spurred to develop a better framework. There is a growing emphasis on embedding so-called ethical principles into AI development and data use. This includes bias detection and mitigation, responsible data-collection practices, and ensuring user control over data. Companies that prioritize transparency and ethical practices are likely to gain consumer trust and loyalty. Of course, the problem with all this regulation is that similar regulation has existed in the past. It is worked around by the largest, monopolistic enterprises, and any fines they usually pay are little more than a slight slap on the wrist, if that. The companies that pay the greatest price are the smaller, more innovative players and, ultimately, their end users.

The benefits of data sovereignty are clear, but there are several challenges that need to be addressed. Dealing with complex data-protection laws across different jurisdictions can be challenging for any organization. Ensuring compliance with these laws requires robust data-governance frameworks and continuous monitoring. Some countries mandate that certain types of data be stored within their borders, which can complicate data management for multinational companies. Solutions include establishing local data centers and implementing data-residency controls to comply with local regulations. Recently, Amazon's AWS cloud business announced a $5.3B investment in Saudi Arabia to build local data centers that can be used to deliver apps and services to nearby customers.

Protecting data from cyber-threats while ensuring privacy is a significant concern. Strong encryption, access controls, and regular security audits are essential to safeguarding data in a sovereign framework. Developing effective and user-friendly models for data monetization is crucial. Platforms like Navigate (nvg8.io) are pioneering these efforts by allowing users to pool and monetize their data collectively, ensuring that the benefits of data are shared more equitably.

The antidote to Big-Tech-controlled-AI-knowing-everything-about-you is AI at the edge, on your own devices. The integration of personal AI with edge computing and federated learning is revolutionizing how we process and use data. These technologies enable AI to operate directly on user devices, enhancing privacy, reducing latency, and allowing for more personalized and responsive AI applications. Edge computing has the potential to protect user data while also bringing computational capabilities closer to data sources, such as smartphones and IoT devices. This proximity reduces the need to transmit data to centralized servers, minimizing latency and enhancing real-time processing capabilities. When combined with AI, edge computing enables devices to execute complex algorithms locally, providing immediate insights and actions based on the data they generate.

Federated learning complements edge computing by enabling the training of AI models across multiple decentralized devices without sharing raw data. Instead, each device trains its model locally and shares model updates with a central server only, which aggregates these updates to create a global model. This approach preserves privacy while benefiting from the collective learning of all participating devices.

Federated learning significantly enhances privacy by keeping raw data on local devices. This reduces the risk of data breaches and ensures that sensitive information remains private. Techniques like secure multiparty computation and differential privacy further protect data during the training process.

One notable example of federated learning in action is Google's Gboard, the company's virtual keyboard for Android and iOS devices. Gboard uses federated learning to improve its predictive text and emoji suggestions without sending users' raw typing data to Google's servers. Instead, each device learns from the user's typing patterns and shares the model updates with the central server only, which then aggregates the updates to enhance the global model. In theory, this approach ensures that users' sensitive typing data remain on their devices while still benefiting from the collective learning of millions of users. The algorithms make it possible, but it's the wielder of the algorithm that many of us distrust. Perhaps it's time to wield such algorithms ourselves.

Advances in edge AI have led to more efficient processing on edge devices, reducing the need for extensive computational resources. By processing data locally, edge AI minimizes the bandwidth required for data transmission and optimizes resource use, making it suitable for real-time applications in areas like health care, the automotive industry, and smart cities. The combination of edge computing and federated learning has enabled new applications across various fields. For instance, in health care, edge AI can process patient data in real time to provide immediate insights and support decision-making. In autonomous vehicles, these technologies enable quick, on-the-spot processing of sensor data, which is crucial for safe and efficient navigation.

Federated learning's decentralized nature allows it to scale across millions of devices, each contributing to and benefiting from the shared global model. This scalability is essential for applications requiring large-scale data processing and collaborative learning across diverse environments and datasets. The benefits are substantial, yet there are challenges associated with federated learning and edge computing. Edge devices often have limited computational power when compared to centralized servers.

Solutions involve optimizing algorithms for efficiency and leveraging hardware advancements to enhance processing capabilities on edge devices. The data across different devices can be highly varied, which poses challenges for model training. Advanced aggregation techniques and adaptive-learning algorithms are being developed to address these issues, ensuring robust and accurate global models.

Federated learning requires frequent communication between devices and central servers. Techniques to reduce communication rounds and optimize data transmission are critical to maintaining efficiency and performance.

The integration of personal AI with edge computing and federated learning is paving the way for a more decentralized, efficient, and privacy-preserving approach to AI. These technologies are transforming various industries by enabling real-time data processing and personalized AI applications while safeguarding user privacy. As advancements continue, the potential for edge AI and federated learning to drive innovation and improve our digital experiences will only grow.

The antidote to ever-faster machines undermining our security and privacy is quantum-safe encryption. As quantum computing continues to advance, it poses a significant threat to current encryption methods. Quantum-safe, or post-quantum, cryptography is essential to protect sensitive data against future quantum attacks. The National Institute of Standards and Technology (NIST) has been at the forefront of developing and standardizing quantum-resistant cryptographic algorithms to ensure the security of our digital information. NIST has announced the first group of quantum-resistant cryptographic algorithms designed to withstand attacks from future quantum computers. These algorithms include CRYSTALS-Kyber, designed for general encryption, valued for its efficiency and relatively small key sizes (the length of the cryptographic key, which impacts both security strength and computational efficiency), making it suitable for secure communications over the internet. CRYSTALS-Dilithium, used for digital signatures, ensures the authenticity and integrity of digital messages and documents.

FALCON, another digital-signature algorithm, provides an alternative to CRYSTALS-Dilithium with different trade-offs in terms of performance and security. SPHINCS+, a stateless hash-based signature scheme, offers strong security assurances and is designed to be highly versatile. (A stateless hash-based signature scheme uses cryptographic functions that transform data into a fixed-size, unique, and unguessable output.)

The adoption of these quantum-safe algorithms is critical for ensuring the long-term security of data. Companies like IBM have already started integrating these algorithms into their products. For instance, IBM's z16 system uses CRYSTALS-Kyber and CRYSTALS-Dilithium for its key-encapsulation and digital-signature functionalities, respectively. This system represents one of the first commercial applications of quantum-safe cryptography. Other tech giants, including Cloudflare and Amazon, are also incorporating quantum-safe algorithms into their security frameworks. Cloudflare's Interoperable, Reusable Cryptographic Library (CIRCL) includes Kyber, and Amazon's AWS Key Management Service supports hybrid modes involving Kyber, providing early adopters with the tools needed to transition to a quantum-safe future.

Although significant progress has been made, transitioning to quantum-safe cryptography presents several challenges. Quantum-safe algorithms often require more computational resources compared to traditional algorithms. Efforts are ongoing to optimize these algorithms to ensure that they can be efficiently implemented across various platforms without significant performance degradation. Ensuring that new quantum-safe cryptographic systems can seamlessly integrate with existing infrastructure is crucial. This includes developing standards and protocols that facilitate smooth transitions and interoperability among different cryptographic systems. Organizations need to be educated about the importance of quantum-safe cryptography and be prepared for the transition. This involves understanding the implications of quantum computing for security and implementing strategies to mitigate potential risks. Quantum-safe encryption is a vital step toward securing our digital future against the threats posed by quantum computing. The advancements made by NIST and the adoption of quantum-resistant algorithms by leading technology companies mark significant milestones in this ongoing effort. As these technologies become standardized and more widely implemented, they will play a crucial role in protecting sensitive data and maintaining the integrity of our digital systems.

The antidote to a global user ID system reminiscent of a dystopian "identity card" scheme is decentralized identity management. Decentralized identity systems represent a transformative approach to managing digital identities by shifting control from centralized authorities to individuals. These systems use technologies like blockchain and cryptographic techniques to enhance security, privacy, and user autonomy. Decentralized identifiers (DIDs) are unique strings that represent entities in the digital realm. These identifiers are stored on blockchain or distributed-ledger technologies, ensuring their immutability and security. DIDs empower users by allowing them to control their digital identities without relying on centralized intermediaries. The World Wide Web Consortium (W3C) has been instrumental in standardizing DIDs to ensure interoperability across various platforms.

Verifiable credentials (VCs) are digital certificates that prove certain attributes or qualifications, similar to physical documents like passports or diplomas. These credentials are crucial for establishing trust in decentralized identity systems. By embedding secure, bidirectional communication features, VCs enable seamless and secure interactions across different systems. In the travel industry, for example, VCs can streamline the verification process for passports, visas, and boarding passes, significantly reducing wait times and errors. Recent developments in decentralized identity systems have focused on improving scalability and performance. Microsoft's Identity Overlay Network (ION), built on the Bitcoin blockchain, is designed to handle tens of thousands of DID operations per second. This scalability is achieved through the Sidetree protocol, which enables efficient management of public key infrastructure (PKI) without compromising decentralization.

Decentralized identity systems have a wide range of applications across various sectors. In health care, they simplify patient-data management while ensuring privacy and security. In finance, they enhance know-your-customer (KYC) processes, reducing fraud and improving customer experience. In travel, they streamline identity-verification processes, making travel more efficient and secure. In retail, decentralized identity systems allow for personalized customer experiences without compromising data security. Despite their potential, decentralized identity systems face several challenges. Ensuring seamless integration across different platforms and services requires standardized protocols and widespread adoption. Figuring out the complex regulatory questions will be essential for widespread adoption. Governments and regulatory bodies need to recognize and support decentralized identity standards.

The antidote to subscription software that you can never own or control is open source. Open-source software (OSS) continues to be a cornerstone of technological innovation and collaboration. By fostering community-driven development, OSS offers robust, flexible, and cost-effective solutions that are widely adopted across industries. In 2024 several key advancements and trends are shaping the landscape of open-source software. Open-source platforms such as TensorFlow and PyTorch are at the forefront of AI and machine-learning development. These tools have made significant strides in enabling developers to build and deploy AI models efficiently.

The integration of AI/ML capabilities into OSS is enhancing productivity and creating new opportunities for innovation across various sectors, from health care to finance.

The importance of security in OSS cannot be overstated. Projects like CycloneDX, which provides a standardized format for software bill of materials (SBOMs), are important in enhancing supply-chain security. CycloneDX v1.6 introduces capabilities for cryptographic attestations and transparency in AI/ML model cards, promoting sustainable and secure software practices. These advancements help organizations ensure compliance with security standards and reduce the risks associated with open-source software deployment.

Progressive web apps (PWAs) are becoming increasingly popular because of their ability to offer a seamless user experience across different devices and platforms. By combining the best features of web and mobile applications, PWAs enhance accessibility and engagement. Major tech companies, including Google, Apple, and Microsoft, are supporting this technology, which is expected to see even broader adoption in the coming years. The integration of security practices into the DevOps pipeline, known as DevSec-Ops, is gaining momentum. This approach ensures that security is a core component of the software-development life cycle, from initial design to deployment. By fostering collaboration among development, operations, and security teams, DevSecOps helps organizations identify and mitigate security vulnerabilities early, leading to more secure software products.

Other "at the edge" computing technologies are reshaping the way that applications are developed and deployed. Kubernetes continues to lead in container orchestration, enabling scalable and efficient management of containerized applications. Edge computing reduces latency by processing data closer to the source, which is essential for applications requiring real-time responses, such as IoT and AI. Python remains a dominant programming language in the open-source community, particularly in AI, machine learning, and data science. Its versatility and extensive libraries make it an ideal choice for developers working on complex projects across various domains. The demand for Python skills is expected to grow, further solidifying its position in the software-development landscape. Open-source software is driving significant advancements in technology and security. By embracing trends like AI integration and enhanced security practices, the open-source community continues to lead innovation.

As these technologies evolve, they will play a critical role in shaping the future of software development and ensuring robust, secure, and efficient digital solutions.

Community-owned infrastructure represents a transformative approach to managing and using resources at the local level. This model empowers communities to take control of essential services and infrastructure, fostering sustainable development and economic growth. For example, blockchain technology is revolutionizing how energy can be generated and traded locally. Projects like EnerPort in Ireland and Brooklyn Microgrid are pioneering peer-to-peer (P2P) energy-trading models. These initiatives enable communities to generate, trade, and consume energy within a decentralized framework, promoting energy independence and sustainability. Such models also facilitate the integration of renewable energy sources and enhance grid stability.

Decentralized autonomous organizations, a technology we have surveyed in this book, are a crucial enabler of freedom, reshaping governance in the cybernetic age. By harnessing the power of blockchain and smart contracts, DAOs create decentralized, transparent, and democratic decision-making processes that have the potential to revolutionize how we structure and manage organizations. The "autonomous" in DAOs is partially about the fact that code executes decisions, but also that it is self-governing.

The concept of interconnected DAOs can open up new possibilities for creating a hierarchy of democratic decisions that span multiple levels of organization. Just as individual DAOs can make decisions through a transparent and decentralized process, networks of DAOs can coordinate and collaborate to address larger-scale issues and challenges. This "DAO of DAOs" model could potentially create a more resilient and adaptive governance system that is better equipped to handle the complexities of the cybernetic era.

In the cybernetic societies of the future, DAOs can drive a variety of critical decisions, ranging from local community projects to global initiatives addressing climate change, public health, and economic stability. By decentralizing decision-making, DAOs enable more inclusive participation, ensuring that diverse perspectives are considered. This inclusive approach not only enhances the legitimacy of decisions but also taps into a broader pool of knowledge and creativity, leading to more innovative and effective solutions.

Real-world examples like MakerDAO and Decentraland showcase the potential of DAOs in action. MakerDAO, one of the most prominent DAOs in the blockchain space, is responsible for managing the Dai stablecoin, a decentralized cryptocurrency that maintains a stable value relative to the US dollar. Through a complex system of smart contracts and community governance, MakerDAO can make critical decisions about the Dai ecosystem, such as adjusting interest rates and collateral requirements, in a transparent and democratic manner. Similarly, Decentraland, a virtual-world platform governed by a DAO, allows users to buy, sell, and develop virtual land using the platform's native cryptocurrency, MANA. The Decentraland DAO enables users to propose and vote on changes to the platform's policies and features, creating a more participatory and community-driven virtual world.

The evolution of DAOs is closely tied to advancements in blockchain technology. Innovations such as layer-2 scaling solutions are crucial for enhancing the efficiency and speed of transactions on the blockchain. Layer-2 solutions work by processing transactions off the main blockchain layer (layer-1) while still leveraging the security of the main blockchain. This approach reduces congestion on the blockchain, lowers transaction fees, and increases throughput, making DAOs more scalable and practical for widespread use.

Additionally, advancements in cryptographic techniques, such as zero-knowledge proofs, offer enhanced privacy and security features. Zero-knowledge proofs allow one party to prove to another that a statement is true without revealing any information beyond the validity of the statement itself. This technology can be crucial for DAOs, ensuring that sensitive information remains confidential while still allowing for transparent and verifiable decision-making processes.

The promise of DAOs as a technology of freedom is clear. By creating transparent, decentralized, and democratic systems for decision-making and resource allocation, DAOs have the potential to fundamentally reshape how we structure and manage organizations in the cybernetic era. As we work to build a more equitable, sustainable, and resilient future, the continued development and adoption of DAOs will be a critical area of focus and innovation.

Embracing the transformative potential of DAOs, we can unlock new frontiers of collaboration, creativity, and resilience, paving the way for a more inclusive and participatory society in the age of cybernetics.

The antidote to expensive energy and a government or giant corporation's control over who can have how much and when is the microgrid. Microgrids are localized energy systems that can operate independently or in conjunction with the main power grid. They are particularly beneficial for rural and remote communities, providing reliable and sustainable energy solutions. Microgrids support local job creation, economic development, and environmental sustainability. Successful projects like the Remote and Indigenous Communities Microgrids Program in Canada demonstrate the potential of microgrids to enhance energy resilience and reduce costs for communities. In the academic and research sectors, community-owned infrastructure is crucial for maintaining control over data and research processes. Organizations like SPARC are advocating for community ownership to counter the trend of commercial acquisition of critical infrastructure. By fostering open access to research and data, these initiatives ensure that educational and research institutions retain control over their resources and can address inequities in access and participation.

Despite their benefits, community-owned infrastructure projects face several challenges. Integration with existing infrastructure and compliance with regulatory standards can be complex. Advanced control systems and standardized communication protocols are essential to ensure interoperability and grid stability. Regulatory reforms are also needed to facilitate the deployment and operation of microgrids and other community-owned systems. Securing funding for community-owned projects can be challenging because of high up-front costs and uncertainties around future energy policies. Innovative financing models, policy incentives, and public–private partnerships are crucial to overcoming these barriers. Governments and financial institutions must recognize the long-term benefits of these projects and provide the necessary support.

Ensuring the long-term sustainability of community-owned infrastructure requires robust planning and management strategies. Initiatives like the "It Takes a Village" project by LYRASIS provide frameworks and tools to help communities sustain open-source software programs and other infrastructure projects. These resources support community engagement, collaborative decision-making, and continuous improvement. Community-owned infra-structure empowers local communities to manage their resources effectively, promoting sustainability, economic growth, and resil-ience. By addressing technical, regulatory, and financial challenges, and leveraging innovative technologies and collaborative models, these initiatives can create a more equitable and sustainable future.

Participatory-governance platforms leverage digital technolo-gies to enhance citizen engagement, transparency, and collaboration in decision-making processes. These platforms are transforming the ways in which governments and citizens interact, making gov-ernance more inclusive and responsive. Digital-participatory plat-forms are being increasingly used to foster citizen engagement and democratic participation. These platforms allow citizens to provide input on policy decisions, participate in budget allocations, and collaborate on community projects. The use of these platforms can disrupt traditional governance structures by enabling more direct and widespread citizen involvement.

Participatory budgeting is a democratic process in which community members directly decide how to spend part of a pub-lic budget. This practice, which originated in Porto Alegre, Brazil, has spread globally and is now being implemented in various cit-ies around the world. Participatory-budgeting platforms facilitate this process by providing tools for citizens to propose, discuss, and vote on budgetary decisions, ensuring that public funds are allocated according to community needs and preferences. Digital platforms enhance transparency and accountability in governance by making information more accessible to the public. These platforms allow for the real-time sharing of data, decision-making processes, and outcomes. For example, platforms like Decidim, used by several municipalities, provide an open-source framework for participatory democracy, enabling transparent and accountable decision-making.

One of the significant challenges in implementing participatory-governance platforms is ensuring that all citizens have access to and can effectively use these digital tools. Efforts to bridge the digital divide include providing digital-literacy programs and ensuring internet access in underserved communities. For digital-participatory platforms to be effective, they must be trusted by the citizens and perceived as legitimate. This requires transparency in how the platforms are managed and how citizen input is used. Building trust also involves ensuring data security and protecting the privacy of participants. Successfully integrating digital-participatory platforms with existing governance structures can be challenging. This integration requires political will, changes in administrative processes, and, sometimes, legislative support to ensure that digital participation is recognized and has a meaningful impact on decision-making.

The future of participatory-governance platforms lies in their ability to adapt and integrate new technologies, such as artificial intelligence and blockchain, to enhance their functionality and security. These advancements can help manage larger volumes of citizen input and provide more robust security measures to protect against cyber-threats. Additionally, fostering partnerships among governments, civic tech organizations, and communities will be crucial in expanding the reach and effectiveness of these platforms. Participatory-governance platforms are reshaping the landscape of democratic engagement by providing new avenues for citizen involvement and transparency. Although challenges remain, the continued development and adoption of these platforms promise a more inclusive and responsive form of governance.

Digital literacy is crucial for navigating the modern world, encompassing a range of skills from basic computer use to advanced problem-solving and critical thinking.

Recent advancements in digital-literacy campaigns highlight the importance of equipping individuals with these skills to foster more inclusive, secure, and innovative societies. Digital-literacy programs are becoming more comprehensive, addressing a wide range of skills necessary for effective digital engagement. For instance, Facebook Philippines' Digital Tayo program has expanded to include lessons on data privacy, cybersecurity, and digital empowerment. This program has successfully reached millions, providing practical lessons through interactive modules and local language support, ensuring that digital literacy is accessible to diverse populations.

The UK's Digital Development Strategy 2024–2030 aims to support digital transformation in developing countries by promoting digital inclusion, responsibility, and sustainability. The strategy emphasizes last-mile connectivity, digital public infrastructure, and artificial intelligence, with a goal of reducing the digital divide by 50 percent in partner countries by 2030. Singapore's efforts to improve digital access and literacy are detailed in the Singapore Digital Society Report. The report highlights the progress made in digital literacy and outlines initiatives to build a more inclusive digital society. These efforts include collaboration among public, private, and people sectors to ensure that digital skills are widely adopted and that residents are prepared for the digital future.

Despite progress, a significant portion of the global population remains offline or lacks basic digital skills. Addressing this requires not only providing hardware but also ensuring access to relevant training and resources. Programs that offer digital-literacy training in local languages and adapt to cultural contexts are essential for reaching underserved communities. As digital engagement increases, so do the risks associated with it, such as cyber-threats and misinformation. Digital-literacy programs must include education on online safety, data privacy, and the ethical use of technology to build a secure and trustworthy digital environment. This is especially critical in developing regions, where digital literacy can directly affect economic and social development.

Sustainable digital-literacy education involves continuous updates to curricula to keep pace with technological advancements. Integrating digital skills into formal education systems and providing lifelong learning opportunities can help maintain a digitally literate population. Collaborative efforts among governments, educational institutions, and private organizations are key to achieving this goal. Digital literacy is foundational to participating in the modern digital economy and society. Recent advancements in digital-literacy campaigns demonstrate a commitment to building inclusive, secure, and resilient digital ecosystems. By addressing challenges and leveraging collaborative efforts, these initiatives can ensure that all individuals are equipped with the necessary skills to thrive in the digital age.

Indeed, the human equation in the age of cybernetics is fundamentally about the interplay among code, consciousness, and control. How human are we as our consciousness evolves with code? How biological are our decisions, in origin, if our skills and mindsets are controlled via code? Code might well turn out to be the medium, with us being both the artists and the art. The adaptability of each individual will be critical in determining whether technology empowers or whether we let it subjugate. One way to adapt is to cultivate fusion skills, to reimagine education and social institutions, and to develop frameworks for human–machine collaboration. Human agency can remain at the center of our cybernetic future.

By embracing these and other technologies of freedom—from decentralized identity systems and open-source software to community-owned infrastructure and participatory-governance platforms—we can begin to chart a new course through the cybernetic landscape. One where technology empowers rather than diminishes us, where innovation serves the interests of the many rather than the few.

But realizing this vision will require more than just individual adoption. It will demand collective action, a social movement committed to building and popularizing the technologies of freedom. We need developers and entrepreneurs to create these tools, investors to fund their development, and ordinary citizens to champion their use.

We need digital-literacy campaigns to equip people with the skills to navigate this new terrain, and we need ethical frameworks to guide the development and deployment of cybernetic technologies. We need policies that incentivize the creation of public goods and rein in the excesses of surveillance capitalism.

Above all, we need a shared vision of a cybernetic future that puts human flourishing at its center—a vision that we can work toward together, one line of code, one community project, and one act of technological self-determination at a time.

The technologies of the cybernetic age are not inherently good or bad. They are tools, and like all tools, their impact depends on how we choose to use them. By embracing the technologies of freedom, we choose to use these tools in the service of human agency, dignity, and empowerment. We choose to be not just consumers of the cybernetic future but also its cocreators.

BEATING CLIMATE CHANGE WITH THE TECHNOLOGIES OF FREEDOM

As we've explored throughout this chapter, the technologies of freedom offer us powerful tools to reclaim our autonomy, enhance our capabilities, and shape a future more in line with the retrofuturistic vision of technology I grew up with. But the potential of these technologies extends far beyond the digital realm. They can also be harnessed to address one of the most pressing challenges of our time: climate change.

This is the hope we have for the MinusFifteen Project, an initiative launched by my wife, Zaib, and me to mitigate the impact of rising temperatures in our hometown of Lahore, Pakistan. The project aims to reduce the city's average summer temperature by 15°F over the next decade through a combination of green infrastructure, community engagement, and data-driven decision-making. Climate change has made the already hot summers unbearable over the past forty years. Increased urban sprawl, a population now in excess of fifteen million people, and the resulting carbon emissions have also contributed.

At the heart of the MinusFifteen Project is a comprehensive monitoring platform that will collect and analyze environmental data from across Lahore. We envision this as a cognitive nexus that learns of and responds to inputs from tens of thousands of sensors. By deploying a vast array of IoT temperature and weather sensors, drones, and satellites, the platform will provide granular insights into the city's urban-heat-island effect, identifying the hottest and coolest areas and suggesting targeted interventions.

But what sets the MinusFifteen Project apart is its commitment to the principles of decentralization, community ownership, and individual empowerment that underpin the technologies of freedom. Rather than relying on centralized infrastructure and top-down control, the project can leverage decentralized mesh networks, such as those being developed by our company SpecFive, to enable the sensors to communicate without the need for paid WiFi or cellular connectivity. This approach not only reduces costs and dependencies but also enhances the resilience and scalability of the monitoring platform. We want to see how far a group of concerned citizens can go in bringing about positive change. What if we don't depend on the government's help to take these important steps for the betterment of the city?

Our plan is that the MinusFifteen Project will actively engage the Lahori community in the data-collection process. Residents will be invited to run temperature sensors in their homes and neighborhoods, contributing to the creation of a detailed heat map of the city. To incentivize participation and reward positive action, we are presently discussing with other early supporters the potential for the project to introduce a blockchain-based token system. With such a reward system in place, in the initial phases participants who run sensors, plant trees, or create green roofs will earn tokens that can be redeemed for products and services provided by supporting companies and individuals, such as paying for a month of cell-phone service or a month of free bus rides.

The potential for this tokenized ecosystem is immense. It creates a new economic model for incentivizing and rewarding climate-positive actions. By earning tokens for their contributions, individuals are not just motivated to participate but also gain a direct economic stake in the success of the project. These tokens can be used to purchase goods and services, driving local economic activity and creating a virtuous cycle of investment in green infrastructure.

By recording these transactions on a blockchain, the system ensures transparency, immutability, and trust. Every action taken, whether it's running a sensor or planting a tree, is securely logged, creating an auditable record of the community's collective impact. This not only helps to quantify the project's success but also provides a powerful narrative tool for inspiring further action. Imagery from drones as well as votes from other participants can be used to validate actions undertaken in the physical world, acting as a sort of "oracle" to ensure that the claims being made regarding contributions are indeed true. If so, rewards are then issued.

As the project evolves, we envision the integration of carbon credits into this tokenized system. Companies or individuals looking to offset their carbon footprint could purchase these tokens, providing a new source of funding for the project and creating a self-sustaining economic model for urban climate action. The value of these tokens would be directly tied to the measurable environmental impact of the project, such as the reduction in urban temperature or the amount of carbon sequestered by new green spaces.

It's still early days, and I definitely don't want to jinx it, but this model has the potential to be very powerful. It aligns economic incentives with environmental goals, empowers individuals to take direct action, and creates a decentralized funding mechanism for sustainable urban development. It's a model that, adapted to local needs and conditions, could be replicated in cities around the world.

Imagine a future where, eventually, the world has its own "green tokens," digital currencies that reward citizens for taking climate-positive actions. These tokens could be earned for anything from using public transport to installing solar panels, from volunteering in community gardens to supporting local green businesses. By gamifying sustainability and creating tangible economic rewards, these systems could drive mass participation in urban climate action.

At the same time, these green tokens could become a means of investment in local sustainable infrastructure. Municipal governments could issue green bonds backed by future token revenues, providing a new way to finance green projects. Impact investors could buy and trade these tokens, creating a new asset class that directly supports sustainable urban development. The more a city reduces its carbon footprint and improves its environmental health, the more valuable its green tokens become, creating a powerful feedback loop of positive change.

If you haven't noted it yet, the MinusFifteen Project can also transform Lahore into a living, breathing cybernetic city. As tens of thousands of sensors proliferate across the urban landscape, communicating with one another through decentralized mesh networks, they form a kind of digital nervous system—a dense web of data flows that provides an unprecedented level of insight into the city's environmental health.

Just as arteries carry oxygenated blood to nourish the body's cells, these data flows will carry vital information about temperature, humidity, air quality, and more. This constant stream of granular, localized data will paint a picture of Lahore's microclimate with a level of detail that has never been possible before. It will allow us to understand the complex interplay of factors that contribute to urban heat islands and to pinpoint the most effective interventions for cooling the city.

And this cybernetic system will also learn and will help the city adapt. Using machine-learning algorithms, the platform will be able to analyze the vast amounts of data it collects, identifying patterns and insights that can guide decision-making. It could, for example, detect that a particular neighborhood is about to experience a heat wave and automatically trigger an alert to local infrastructure and health services. These alerts could result in an increased readiness of the electric grid, a heightened availability of health professionals, or even interventions such as the activation of misting systems.

On a more long-term basis, the MinusFifteen cybernetic core watching over Lahore could identify areas where tree cover is most needed and then direct tree-planting efforts accordingly. It might adjust the incentives offered through the token system in increasing rewards for water conservation during a drought or for reflective roof installation in a heat-prone area.

If this vision succeeds, over the next decade or two, Lahore—one of the world's largest urban centers—can become a city that can sense, respond, and adapt to its environment in real time. It will be a city that is not just smart but also partially autonomous—a city that can learn from its own data and optimize its own performance. And at the heart of this intelligence will be the collective actions of thousands of citizens and tens of thousands of sensors, each contributing their data, their validatory input, and their energies to reach a common goal.

This is the bright side of the cybernetic tomorrow. It's a future where technology empowers communities to take collective action, where data drive decision-making, and where economic incentives are aligned with environmental sustainability. It's a future where cities are not just places to live but are also active participants in the fight against climate change. And in order to succeed it will require a fundamental shift in how we think about urban governance, civic participation, and economic-value cre-ation. It will demand new partnerships among the public sector, the private sector, and civil society. And it will require a willing-ness to experiment, to take risks, and to learn from failure. But in the end, it's all worth it. This is what technology was meant for. To improve our lot.

CHAPTER 10

CYBERNETIC SYNTHESIS

Through the course of this book, we've navigated the complex landscape of human–machine fusion, witnessing the profound ways that this technological revolution is transforming our world. We've taken a look at the various trends shaping the future of human–machine fusion, examining principles like the laws of scaling and patterns of cliodynamics that offer insight into the trajectory of this cybernetic era. We've also talked about a number of well-developed frameworks through which we can predict where things might go next. In this chapter, let's take a deeper view and synthesize all the aspects of our discussion so far.

The very first thing to consider is that we are living in a reflexive world where human intent is magnified by machines and is then translated to action. Central to this transformation is a fundamental shift in the nature of causality where human intentions, amplified and accelerated by machines, can quickly translate into far-reaching actions. That action, if it occurs at sufficient scale and speed, can often make the outcomes it presupposes actually happen. Digital action can become action in the real world, often even driving positive or negative spirals.

This phenomenon of digital actions translating into real-world consequences is particularly evident in the realm of financial markets. High-frequency trading algorithms, which can execute trades in fractions of a second based on complex mathematical models, have the potential to cause significant market movements and even flash crashes. In the infamous 2010 flash crash, the Dow Jones Industrial Average plunged nearly a thousand points in just a few minutes, largely as a result of the action of algorithmic trading systems. This event highlighted the power of machine-driven decision-making to rapidly and dramatically affect the real world.

Another area where this phenomenon is particularly obvious now is with social media. The impact of social media on public opinion and behavior exemplifies this phenomenon. As algorithms shape our information feeds, promoting content that aligns with our preexisting beliefs and emotions, they can subtly influence our perceptions and choices. The number and sophistication of bots expressing an opinion can change the tone of a discussion and indeed the mood of an audience.

A notorious experiment conducted by Facebook researchers in 2014 starkly illustrated the power of social media to manipulate emotions on a massive scale. For one week, Facebook altered the news feeds of 689,003 users, showing some of them more positive content and others more negative content. Facebook then analyzed more than three million posts and found that when exposed to more negative content, people produced more negative posts, and vice versa with positive content. This so-called "emotional contagion" occurred without users' awareness. The experiment demonstrated how even small tweaks to algorithms can have significant effects on the emotional states of huge numbers of people.

I think back to this experiment, where a number of user feeds were changed to promote depressing, negative material without users knowing. In other words, whether or not we actually get depressed can be triggered through the actions of a machine, which may amplify the intent of a human being (or, in this case, of a group of people at Facebook). It's a stark example of how the architecture of online platforms can be used to shape human behavior on a large scale.

The implications of this emotional manipulation extend far beyond individual well-being. Research has shown that emotional states can influence political beliefs, voting behavior, and even economic decisions. A study published in *Cognitive, Affective, & Behavioral Neuroscience* found that inducing a negative mood in participants made them more likely to support conservative political views. Another study, in the *Journal of Consumer Psychology*, demonstrated that people in a sad mood were willing to pay more for products than those in a neutral mood. Therefore, the ability of social media platforms to manipulate emotions on a mass scale has significant societal and political ramifications.

There is much research on how when something starts to seem inevitable, you can cause people to give up and not even try. A study published in the *Journal of Abnormal Psychology* found that people who were conditioned to believe that an unpleasant outcome was inevitable ended up exerting less effort to prevent it. Thus, messaging that something is inevitable will definitely cause changes in behavior. Amplification of human intent through cybernetic systems will cause real behaviors to change; there is no question about this. The corollary is that cybernetic systems—human intent amplified by machines such as social networks—can prime entire societies toward inaction. Don't vote for candidate X because a loss is inevitable.

We also talked about privacy and how with the vast collection of information currently occurring throughout the internet and by intelligence agencies, advertisers, communications companies, and pretty much all businesses, there is little reasonable hope that anyone can maintain their privacy. Amid this shifting terrain, the erosion of privacy emerges as a defining challenge of our time. With every digital interaction we generate a vast trail of personal data, ripe for exploitation by advertisers, governments, and tech giants.

The implications of this pervasive data collection are profound. It enables the creation of incredibly detailed profiles of individuals, revealing their habits, preferences, and vulnerabilities. This information can be used to manipulate consumer behavior, as demonstrated by the Cambridge Analytica scandal, where the personal data of millions of Facebook users were harvested without their consent and used for political advertising. It can also be used for more nefarious purposes, such as stalking, identity theft, or blackmail. In the hands of authoritarian governments, these data could be used to surveil and control populations, stifling dissent and reinforcing power structures.

As the pervasive surveillance apparatus expands, the idea of opting out seems increasingly unrealistic. Laws like Europe's GDPR and California's CCPA aim to give users more control, but their impact remains limited as data collection and sharing form the core business model of many tech giants. True privacy may require a more fundamental rethinking of our relationship with technology companies.

So what do we do in an environment of the type we've created, in concert with technology? Do we accept this as a fait accompli, or do we counter these cybernetic systems with our own? I can't predict what will actually happen, but I can share with you that the choice I am making is to run even faster toward technology so that I can own it and not be controlled by it. This is the alternative approach. Rather than retreating from the technological wave, we might choose to dive deeply into its currents, aiming to harness the tools of the cybernetic age.

Instead of relying on cloud services to store and index my information, I am investing in my own storage systems, taking inspiration from projects like the personal data stores being developed by Sir Tim Berners-Lee's company Inrupt. Instead of just communicating through centralized networks, I am getting into decentralized, pccr-to-peer mesh networks, which we covered in chapters 4 and 8.

Decentralized-storage solutions like the InterPlanetary File System (IPFS) and Filecoin are also gaining traction as alternatives to centralized cloud storage. These systems distribute data across a network of nodes, making it more resilient and resistant to censorship. They also often incorporate built-in economic incentives for nodes to store and maintain data, creating a self-sustaining ecosystem. By combining personal data stores with decentralized storage networks, we can start to envision a future where individuals have much greater control over their own data.

Efforts like NYC Mesh are building community-owned networks that don't rely on big ISPs. Reticulum, Meshtastic, and SpecFive's Hypermesh are other examples of protocols that can be used to build community-enabled communications. With a simple hundred-dollar router, you can join this network and communicate with your neighbors without going through the commercial internet. Quite easily, you can begin to patch together neighborhoods to build a metro-scale system. Extrapolate this model out, and we can envision a future where local communities own at least one type of communications infrastructure, making mass surveillance much harder.

Encryption to keep our data safe and private is already available to us, and as we have seen, even quantum-safe encryption algorithms are now within our reach.

I intend to take advantage of these technologies to keep information safe on systems that I own. Think of all of these as technologies of freedom—things that free us from control, exploitation, and constant observation. We don't have to just accept the erosion of privacy as inevitable. We can actively work to create an alternative technological infrastructure that better protects our human rights and autonomy.

The embrace of technology propels us not just toward technology but also toward the people who are using it in the same way. The interest groups I have chanced upon as I've explored how to develop, use, and implement technologies of freedom have been a joy to get to know. Whether it's the vibrant open-source-software community, the passionate advocates of the decentralized web, or the privacy-focused crypto enthusiasts, there are many people working to build a more empowering technological future.

These are the people exploring how we can harness advanced technologies while protecting our core values. They can and will play the role of teachers, friends, and guides for many of us as we investigate technologies that can protect us and our independence. Learning from and collaborating with these communities will be essential as we navigate the challenges ahead.

The decisions don't merely concern us and our abilities; our brain will eventually come to see the person of "us" as a combination of mechatronics, computing, and biology. The decisions it will then make, aided by data and computation that can be tapped into perhaps via neural interfaces, will be quite different from the types of decisions it makes today.

This fusion of biological and artificial intelligence raises profound questions about the nature of the self. If your thoughts and memories are augmented by digital systems, where does the "you" end and the machine begin? Philosophers and ethicists are forever grappling with these questions. Some, like transhumanist thinker Ray Kurzweil, see this merger as the next stage in human evolution, a chance to transcend our biological limitations. Others, like bioethicist Wesley J. Smith, warn of the dehumanizing potential of these technologies, arguing that they could erode our sense of identity and human dignity.

So as we change ourselves with technology, we change our decisions and actions, and our decisions and actions change the world. This is the cybernetic impact expanding out from the centroid of a single human, rippling, echoing, and affecting the globe.

But much as these changes will affect our personal abilities and decision-making, in the cybernetic society of tomorrow we will also see cities and nation-states change.

This dynamic could play out on a global scale, with nations and cities that are quick to adopt and integrate cybernetic technologies gaining a significant competitive advantage over those that lag behind. Even as we see the emergence of "smart cities" that leverage the collective intelligence of their augmented citizens to solve complex problems and drive innovation, we could see the rise of "ghost cities" that fail to keep pace and are left behind, their populations increasingly marginalized and disconnected from the benefits of the cybernetic revolution.

On the other hand, widespread access to augmentation technologies could have a democratizing effect. If everyone has access to enhanced memory, information processing, and problem-solving skills, it could level the playing field and create a more meritocratic society. Will living in a world of cybernetic cities serve us so well that it will effectively disarm us and, over time, cure us of the need to cling to high positions in a hierarchy as the only way to seek self-satisfaction and self-confirmation?

Much will depend on how these technologies are developed and distributed.

Think back to the cult sci-fi TV show *Fringe*, which painted a vivid picture of a possible human future, one in which humans have elected to grow their cognitive capabilities in a way that transforms them into rather cold prediction machines. Now make the colors on the set of *Fringe* a bit brighter and discard the dystopian implications; the imagined capability is pretty cool! With vast digital memory at our fingertips and AI-powered predictive analytics humming in the background, we may gain an almost superhuman ability to learn from the past and anticipate the future. Every choice could be informed by a comprehensive analysis of its potential consequences.

This predictive capability could fundamentally change how we approach decision-making. Instead of relying on intuition or limited information, we could run detailed simulations of different scenarios, testing out the long-term implications of our choices. This could lead to more informed, strategic decisions at every level, from personal life choices to corporate strategy to government policy. However, such a future raises questions about the role of human judgment and intuition. If we become too reliant on predictive algorithms, could we lose our ability to think creatively and adapt to unexpected situations? Or can algorithms make us more creative than we can be on our own? I think they can.

Of course, such powerful predictive abilities will inevitably be applied to warfare as well. We will have to confront the automation of war and the fact that small numbers of humans will use the cybernetic infrastructure of war to have an outsized impact on world affairs. As we've seen in the chapter on hyperwar, the US military is already investing heavily in projects like the "Third Offset" strategy, which seeks to maintain American military superiority through the integration of AI, robotics, and other advanced technologies. China and Russia are not far behind.

Smaller countries will indeed be more influential and powerful based on their mastery of these technologies, as military strategists have predicted based on current trends. A report by the Center for a New American Security argues that AI will be a strategic equalizer, allowing smaller states and even nonstate actors to compete with larger powers. Autonomous weapons, swarm tactics, and cybernetic enhancement of soldiers could change the balance of power and reshape global geopolitics.

This democratization of military power could have destabilizing effects. In a world where a small group of enhanced individuals can wield nation-level destructive capabilities, the risk of terrorism, insurgency, and asymmetric warfare could increase dramatically. At the same time, the prospect of mutually assured destruction (MAD) might become even more pronounced because the barriers to entry for devastating attacks are lowered. Managing this risk will require new forms of international cooperation and governance to regulate the development and use of these technologies.

Will this mean a future where a Dr. Doom–like figure can become a major force to reckon with as a result of the technology at his disposal? Or a world where people learn to live together in greater peace and harmony because all nations might realize that the use of autonomy and AI, coupled with a very small number of human beings, results in a potent capacity for destruction: a kind of democratization of mutually assured destruction. Nuclear weapons granted this terrible capacity to a small handful of superpowers. But AI and other exponential technologies may spread this MADness to a point where we are cured. After all, if many of the countries we deal with can press the proverbial button, perhaps we will be compelled to find ways to get along.

As with all powerful technologies, much will depend on the wisdom with which we wield them. There will no doubt be those who will act as anarchists and in whose interest it is to upset any new balance of world order. Nonstate actors empowered by cybernetic technologies could become major destabilizing forces, pushing against the established power structures. We will see who succeeds in this new landscape of power and conflict.

This is where the role of international institutions and global governance will be critical. Just as the nuclear age required the development of new treaties, international laws, and monitoring systems, the cybernetic age will demand new forms of global cooperation to manage the risks and benefits of these technologies. If this involves regulations that prohibit certain technologies from being developed, or sanctions or other restrictive approaches, then I fear that these attempts will fail. If they help individuals use technology for real, near-term benefit, then they will succeed.

Ultimately, the path forward will require reflection and a willingness to learn from the past in terms of the relationship between technology and progress on the one hand, and regulation and technology on the other. We will need to develop new frameworks for thinking about the rights and responsibilities of cognitively enhanced individuals and the societies they inhabit. We will need to grapple with profound questions about the nature of identity, agency, and being during an age of increasing symbiosis with technology. Most importantly, we will need to cultivate the moral courage to use these powerful technologies for the greater good, resisting the temptations of short-term gain or destructive power.

In this ongoing synthesis of human and machine, of biology and technology, lies the promise of the next stage of our evolution and the peril of our demise. My hope is that we can rise to meet this moment with open minds, compassionate hearts, and a commitment to the betterment of ourselves and our world. The cybernetic frontier stretches out before us, rich with possibility—but it falls to us to explore its uncharted territory, guided by the compass of a noble conscience. That is the future I hope for.

ACKNOWLEDGMENTS

My undying gratitude to my mother for her encouragement and her belief in me. Her prayers have forever been the wind in my sails.

To my beloved wife, Zaib—your love is my anchor, your calm presence a source of strength. Thank you for your patience and the countless ways you've made writing this book easier.

To my sons, Asas, Murtaza, and Hyder—thank you for humoring me, reading through the manuscript, and sharing your thoughtful insights. They were invaluable in shaping what this book has become.

To my sister, Tasneem Zehra—thank you for reviewing the manuscript and encouraging me during those long hours in Boston as I raced to meet my editorial deadline before heading to the Middle East.

To Ali Husain and Mahe Zehra—my younger siblings, for always being there for me and for their constant love, trust, and support. And to my siblings-in-law, Abdullah, Saad, and Zehra.

To my wonderful editor, TJ Kelleher at Basic Books—who certainly gave me a lot of work to do! But whose every suggestion, every tough edit, made this book so much better. I am deeply grateful for your dedication and your craft.

To all of you, my deepest gratitude.

FURTHER READING
AND REFERENCES

INTRODUCTION: CODE, CONSCIOUSNESS, AND CONTROL

The sentient machine: Amir Husain, *The Sentient Machine* (Scribner, 2017).

Marc Andreessen's concept of "software eating the world": Marc Andreesen, "Why Software Is Eating the World," Andreessen Horowitz, 2011, https://a16z.com/why-software-is-eating-the-world.

Norbert Wiener's definition of *cybernetics*: Norbert Wiener, *Cybernetics or Control and Communication in the Animal and the Machine* (MIT, 1948).

Norbert Wiener's life: "Norbert Wiener, American Mathematician," Britannica, www.britannica.com/biography/Norbert-Wiener.

Annual data generation per person on Earth by 2025: more than 20 terabytes: Aditya Rayaprolu, "25+ Impressive Big Data Statistics for 2024," TechJury, January 3, 2024, https://techjury.net/blog/big-data-statistics.

George Soros's concept of reflexivity: George Soros, *The Alchemy of Finance* (Simon & Schuster, 1987).

Brookings Institution study on US workers facing high exposure to automation: Mark Muro, Jacob Whiton, and Robert Maxim, "What Jobs Are Affected by AI?," Metropolitan Policy Program at Brookings, 2019, www.brookings.edu/wp-content/uploads/2019/11/2019.11.20_Brookings Metro_What-jobs-are-affected-by-AI_Report_Muro-Whiton-Maxim .pdf.

Pakistan solar power: Faseeh Mangi, "Pakistan Sees Solar Boom as Chinese Imports Surge, BNEF Says," Bloomberg, August 9, 2024, www.bloomberg .com/news/articles/2024-08-09/pakistan-sees-solar-boom-as-chinese -imports-surge-bnef-says?.

World Economic Forum report on ethical challenges in the fourth industrial revolution: Beena Ammanath, Kay Firth-Butterfield, and Don Heider, "Ethics by Design: An Organizational Approach to Responsible Use of Technology," World Economic Forum, December 2020.

UN Secretary-General's report: UN Secretary-General's High-Level Panel on Digital Cooperation, "The Age of Digital Interdependence," United Nations, 2019.

Pew survey on perceptions of AI: Michelle Faverio and Alec Tyson, "What the Data Says About Americans' Views of Artificial Intelligence," Pew Research Center, www.pewresearch.org/short-reads/2023/11/21/what-the -data-says-about-americans-views-of-artificial-intelligence.

CHAPTER 1: THE ORIGINS OF CYBERNETICS

Norbert Wiener and Arturo Rosenblueth's work on cybernetics in the 1930s: Norbert Wiener, *Norbert Wiener—A Life in Cybernetics* (MIT Press, 2018), https://ieeexplore.ieee.org/servlet/opac?bknumber=8327688.

André-Marie Ampère's concept of "cybernétique" in the nineteenth century: "André-Marie Ampère," *Physics Today*, January 20, 2015, https://pubs.aip .org/physicstoday/Online/8289/Andre-Marie-Ampere.

Wiener's book defining *cybernetics*: Norbert Wiener, *Cybernetics or Control and Communication in the Animal and the Machine* (MIT, 1948).

Dartmouth conference on artificial intelligence (1956): "Dartmouth Workshop," Wikipedia, retrieved July 2024, https://en.wikipedia.org/wiki /Dartmouth_workshop.

Marvin Minsky's development of the perceptron: Marvin Minsky and Seymour Papter, *Perceptrons: An Introduction to Computational Geometry* (MIT Press, 1969).

John McCarthy's development of the Lisp programming language: "Lisp (Programming Language)," Wikipedia, retrieved July 2024, https://en .wikipedia.org/wiki/Lisp_(programming_language)#History.

Daniel Kahneman, *Thinking, Fast and Slow* (Farrar, Straus and Giroux, 2013).

Erik Eyster, Shengwu Li, and Sarah Ridout, "A Theory of *Ex Post* Rationalization," https://economics.harvard.edu/files/econ/files/li_fall_2021.pdf.

On error-correcting codes and adaptive programming techniques: www .researchgate.net/publication/366814970_Methods_to_Realize_Low -BER_and_High-Reliability_RRAM_Chip_With_Fast_Page-Forming _Capability.

CHAPTER 2: NEOM AND THE WORLD OF THE FUTURE

Neom's estimated population: Paige Peterson, "NEOM—The Line," National Council on US-Arab Relations, September 5, 2023, https://ncusar.org/blog/2023/09/neom-the-line.

Estimated cost of building Neom: Fahad Abuliadayel, "Saudi Arabia's City of the Future Gets $5.6 Billion Investment," Bloomberg.com, June 6, 2023, www.bloomberg.com/news/articles/2023-06-06/saudi-arabia-s-city-of-the-future-gets-5-6-billion-investment.

Saudi Arabia's Vision 2030 initiative: "Saudi Vision 2030: A Story of Transformation," www.vision2030.gov.sa/en.

Neom's planned area: Merlyn Thomas and Vibeke Venema, "Neom: What's the Green Truth Behind a Planned Eco-City in the Saudi Desert?," *BBC News*, February 21, 2022, www.bbc.com/news/blogs-trending-59601335.

Concept of the "cognitive city" in Neom: "The Line," Neom.com, www.neom.com/en-us/regions/theline.

Planned features of Neom's transportation system: "The Future of Mobility," Neom.com, www.neom.com/en-us/our-business/sectors/mobility.

Neom's commitment to renewable energy and sustainable infrastructure: "Saudi Arabia Mega City Neom to Run Entirely on Renewable Energy," CNBCTV18.com, February 20, 2023, www.cnbctv18.com/world/saudi-arabia-mega-city-neom-to-run-entirely-on-renewable-energy-15979881.htm.

Planned AI-assisted urban planning and design in Neom: "The Future of Urban Design: AI in NEOM's City Planning," Cyberfutures.com, https://cyberfutures.ai/the-future-of-urban-design-ai-in-neoms-city-planning.

Ethical and political questions raised by the integration of AI into urban governance and decision-making: Wenjing Zhu, "Artificial Intelligence and Urban Governance: Risk Conflict and Strategy Choice," *Open Journal of Social Sciences* 9, no. 4 (April 2021), www.scirp.org/journal/paperinformation?paperid=108627.

CHAPTER 3: COMPANIES AS CYBERNETIC ORGANISMS

Nvidia and flat organizations: Derrick Clinton, "NVIDIA's Meteoric Rise Fueled by Flat Organizational Structure and Innovative Workspace Design," *Cryptopolitan*, November 19, 2023, www.cryptopolitan.com/nvidias-rise-fueled-by-flat-organizational.

AI and infrastructure: Laureano Alvarez and Galo De Reyna, "The Age of With . . . AI in Construction and Infrastructure," Deloitte.com, April 27, 2020, www.deloitte.com/ce/en/industries/industrial-construction/perspectives/the-age-of-with-ai-in-construction-and-infrastructure.html. For the McKinsey report, see Angela Spatharou, Solveigh Hieronimus, and Jonathan Jenkins, "Transforming Healthcare with AI," McKinsey, March 10, 2020, www.mckinsey.com/industries/healthcare/our-insights/transforming-healthcare-with-ai.

Bezos: Michael Schrage, "R&D, Meet E&S (Experiment & Scale)," *MIT Sloan Management Review*, May 11, 2016; Michael Shick, "Jeff Bezos: What's Dangerous Is Not to Evolve," *Fast Company*, March 3, 2010, www.fastcompany.com/1569357/jeff-bezos-whats-dangerous-not-evolve.

Amazon and robots: Sebastian Herrera, "Amazon Introducing Warehouse Overhaul with Robotics to Speed Deliveries," *Wall Street Journal*, October 18, 2023; Caleb Naysmith, "Amazon Grows to Over 750,000 Robots as World's Second-Largest Private Employer Replaces Over 100,000 Humans," *Benzinga*, April 11, 2024, www.benzinga.com/general/24/04/38202793/amazon-grows-to-over-750-000-robots-as-worlds-second-largest-private-employer-replaces-over-100-000.

Amazon's apprenticeship program on robotics: Connie Chen, "An Amazon Apprenticeship Program Pays Employees to Gain Skills and Learn About Robotics," Aboutamazon.com, May 4, 2023, www.aboutamazon.com/news/operations/amazon-pays-employees-to-gain-skills-in-robotics.

Amazon Robotics: "Amazon Robotics," Wikipedia, retrieved September 2024, https://en.wikipedia.org/wiki/Amazon_Robotics; "The Story Behind Amazon's Next Generation Robot," Aboutamazon.com, March 11, 2019, www.aboutamazon.com/news/innovation-at-amazon/the-story-behind-amazons-next-generation-robot; Will Knight, "Amazon's New Robots Are Rolling Out an Automation Revolution," *Wired*, June 26, 2023, www.wired.com/story/amazons-new-robots-automation-revolution.

Self-driving warehouse cart: Brian Heater, "Amazon Acquires Autonomous Warehouse Robotics Startup Canvas Technology," TechCrunch, https://techcrunch.com/2019/04/10/amazon-acquires-autonomous-warehouse-robotics-startup-canvas-technology.

Agility Robotics and Ford: "Agility Robotics to Sell First Digit Robots to Ford to Accelerate Exploration of Commercial Vehicle Customer Applications," Ford.com, January 6, 2020, https://media.ford.com/content

/fordmedia/fna/us/en/news/2020/01/06/agility-robotics-sell-first-digit
-robots-to-ford.html.

Agility Robotics and Amazon: Sebastian Herrera, "Amazon Introducing Warehouse Overhaul with Robotics to Speed Deliveries," *Wall Street Journal*, October 18, 2023.

Amazon, productivity tracking, and firing, including of Stephen Normandin: Colin Lecher, "How Amazon Automatically Tracks and Fires Warehouse Workers for 'Productivity,'" *Verge*, April 25, 2019; Spencer Soper, "Fired by Bot at Amazon," Bloomberg, June 28, 2021, www.bloomberg.com /news/features/2021-06-28/fired-by-bot-amazon-turns-to-machine -managers-and-workers-are-losing-out.

Hands off the Wheel: Spencer Soper, "Amazon Began Automating Warehouses a While Ago: Now Its Machines Get Desk Jobs Too," *Los Angeles Times*, June 13, 2018, www.latimes.com/business/la-fi-amazon-automation-jobs -20180613-story.html; Alex Kantrowitz, "How Amazon Automated Work and Put Its People to Better Use," *Harvard Business Review*, September 16, 2020, https://hbr.org/2020/09/how-amazon-automated-work-and-put-its -people-to-better-use.

French fine: Sam Gruet, "Amazon Fined for 'Excessive' Surveillance of Workers," *BBC News*, January 23, 2024, www.bbc.com/news/business -68067022.

Estimated improvement in profitability for companies developing responsible AI systems by 2035: "The Art of AI Maturity," Accenture, 2017, www .accenture.com/us-en/insights/artificial-intelligence/ai-maturity-and -transformation.

Geoffrey West's book: Geoffrey West, *Scale: The Universal Laws of Life, Growth and Death in Organisms, Cities, and Companies* (Penguin, 2017).

Philip Zimbardo's car study: Philip G. Zimbardo, "The Human Choice: Individuation, Reason, and Order Versus Deindividuation, Impulse, and Chaos," Nebraska Symposium on Motivation, 1969, https://stacks.stanford .edu/file/gk002bt7757/gk002bt7757.pdf.

George L. Kelling and James Q. Wilson, "Broken Windows: The Police and Public Safety," *The Atlantic*, March 1982, www.theatlantic.com/magazine /archive/1982/03/broken-windows/304465.

CHAPTER 4: A WORLD OF NEOMS

Saudi development plans: "Vision 2023," www.vision2030.gov.sa.

United Arab Emirates: Shaista Khan, "How District 2020 Is the Future of Urban Planning and Development," *Fast Company Middle East*, April 18, 2022.

Iridium satellite project by Motorola: Sydney Finkelstein and Shade H. Sanford, "Learning from Corporate Mistakes: The Rise and Fall of Iridium," Dartmouth Tuck School of Business, https://mba.tuck.dartmouth.edu /pages/faculty/syd.finkelstein/articles/Iridium.pdf.

George Gilder's work: George Gilder, *Knowledge and Power: The Information Theory of Capitalism and How It Is Revolutionizing Our World* (Regnery, 2013).

Claude Shannon's information theory: "Information," *Stanford Encyclopedia of Philosophy*, https://plato.stanford.edu/entries/information.

SparkCognition: "Press Releases," Sparkcognition, accessed September 27, 2024, www.sparkcognition.com/resources-press-releases.

Mesh networks: "Introduction," Meshtastic, https://meshtastic.org/docs /introduction; "Mesh Networking," Wikipedia, accessed September 2024, https://en.wikipedia.org/wiki/Mesh_networking.

CHAPTER 5: HUMAN AUGMENTATION

Smartphones and thumbs: Anne-Dominique Gindrat et al., "Use-Dependent Cortical Processing from Fingertips in Touchscreen Phone Users," *Current Biology* 25, no. 1 (January 5, 2015): 109–116, www.cell.com/current -biology/fulltext/S0960-9822(14)01487-0; Sophie Evans, "Do You Have This Problem? Bizarre Reason Some of Us Now Have One Thumb BIGGER Than the Other," *Mirror*, June 8, 2016, www.mirror.co.uk/news/uk-news /you-problem-bizarre-reason-now-8140638.

Daugherty and Wilson on skills: Paul R. Daugherty and H. James Wilson, *Human + Machine: Reimagining Work in the Age of AI* (Harvard Business Review Press, 2018).

Caton, Beck, and Berger: Anton Coenen and Oksana Zayachkivska, "Adolf Beck: A Pioneer in Electroencephalography in Between Richard Caton and Hans Berger," *Advances in Cognitive Psychology*, December 31, 2013, https://doi.org/10.2478/v10053-008-0148-3.

Wilder Penfield and Herbert Jasper: William Feindel, "Development of Surgical Therapy of Epilepsy at the Montreal Neurological Institute," *Canadian Journal of Neurological Sciences* 18, no. S4 (November 1991): 549–553, https://doi.org/10.1017/S0317167100032674.

Research on BCIs for treating depression, epilepsy, and Parkinson's disease: Xiaoke Chai et al., "Brain–Computer Interface Digital Prescription for Neurological Disorders," *CNS Neuroscience and Therapeutics* 30, no. 2 (February 2024): e14615, www.ncbi.nlm.nih.gov/pmc/articles /PMC10867871.

Decoding imagined speech: www.biorxiv.org/content/10.1101/2022.08.02 .502503v1.full.

BCIs: Emma Woollacott, "How to Control a Machine Using Your Mind," *BBC News*, February 1, 2018.

Neuralink: Bill Chappell, "What to Know About Elon Musk's Neuralink, Which Put an Implant into a Human Brain," NPR, January 30, 2024, www.npr.org/2024/01/30/1227850900/elon-musk-neuralink-implant -clinical-trial.

Information on Neuralink's projects can be found at http://neuralink.com.

Kernel: www.kernel.com.

Paradromics: Bruce Gil, "A Neuralink Rival Wants to Give People Who Can't Talk Their Voices Back with a High-Tech Brain Chip," *Quartz*, July 28, 2024, https://qz.com/elon-musk-neuralink-paradromic-brain-chip -1851604324; see also www.paradromics.com.

CTRL-Labs wristbands: "Facebook Buys 'Mind-Reading Wristband' Firm CTRL-Labs," *BBC News*, September 24, 2019, www.bbc.com/news /technology-49812689.

Neurable: Emily Mullin, "I Tried These Brain-Tracking Headphones That Claim to Improve Focus," *Wired*, September 24, 2024, www.wired.com /story/this-brain-tracking-device-wants-to-help-you-work-smarter; see also www.neurable.com.

NextMind: Scott Stein, "Mind Control Comes to VR, Letting Me Explode Alien Heads with a Thought," CNET, January 30, 2021, www.cnet.com /tech/computing/controlling-vr-with-my-mind-nextminds-dev-kit -shows-me-a-strange-new-world.

BLEEX (Berkeley Lower Extremity Exoskeleton) project at the University of California, Berkeley: https://bleex.me.berkeley.edu/project/bleex. See also https://pmc.ncbi.nlm.nih.gov/articles/PMC4236484.

Tanya Jonker's research: Tanya Jonker et al., "Neural Reactivation in Parietal Cortex Enhances Memory for Episodically Linked Information," *Proceedings of the National Academy of Science* 115, no. 43, www.pnas.org /doi/abs/10.1073/pnas.1800006115.

Changing a person's understanding of math: www.cell.com/current-biology
/fulltext/S0960-9822%2810%2901234-0?switch=standard.

Robin Dunbar and colleagues on active social-network size and the orbital
prefrontal cortex: https://pubmed.ncbi.nlm.nih.gov/22298855.

Robin Dunbar's concept of "Dunbar's number": Robin Dunbar, "Dunbar's
Number," NewScientist.com, retrieved July 2024.

F. Yu, A. Moehring, O. Banerjee, T. Salz, N. Agarwal, and P. Rajpurkar, "Heterogeneity and Predictors of the Effects of AI Assistance on Radiologists,"
Nature Medicine 30 (2024): 837–849, www.nature.com/articles/s41591
-024-02850-w.

CHAPTER 6: CYBERNETIC CONFLICT: HYPERWAR

History of DARPA's "Third Offset Strategy," calling for leveraging AI and
autonomy to maintain US military technological edge: Gian Gentile et
al., "A History of the Third Offset, 2014–2018," Rand Corporation, March
31, 2021, www.rand.org/pubs/research_reports/RRA454-1.html.

Autonomous weapons as "overhyped": https://scholarlycommons.law.case.edu
/cgi/viewcontent.cgi?params=%2Fcontext%2Fjil%2Farticle%2F1005
%2F&path_info=47CaseWResIntlL2.Article.Noone_26Noone.Print
.pdf.

DoD's directive on autonomous weapons mandating "appropriate levels of
human judgment over the use of force": "DOD Directive 3000.09:
Autonomy in Weapon Systems," US Department of Defense, Office of
the Under Secretary of Defense for Policy, January 25, 2023; Resolution
78/241, "Lethal Autonomous Weapons Systems," adopted by the United
Nations General Assembly on 22 December 2023, Submission of the
United States of America," https://docs-library.unoda.org/General
_Assembly_First_Committee_-Seventy-Ninth_session_(2024)/78-241
-US-EN.pdf.

DoD's "Ethical Principles for Artificial Intelligence": "DOD Adopts Ethical
Principles for Artificial Intelligence," US Department of Defense, February 24, 2020, www.defense.gov/News/Releases/Release/Article/2091996
/dod-adopts-ethical-principles-for-artificial-intelligence.

DARPA's Offensive Swarm-Enabled Tactics (OFFSET) program for developing swarms of collaborative autonomous aircraft: DARPA, "OFFensive
Swarm-Enabled Tactics (OFFSET)," accessed September 27, 2024, www
.darpa.mil/program/offensive-swarm-enabled-tactics.

DARPA's Squad X: "Squad X Program Envisions Dismounted Infantry Squads of the Future," DARPA, March 15, 2016, www.darpa.mil/news -events/2016-03-15.

China's "New Generation Artificial Intelligence Development Plan": Graham Webster et al., "Full Translation: China's 'New Generation Artificial Intelligence Development Plan,'" *Digichina*, August 1, 2017, https://digichina .stanford.edu/work/full-translation-chinas-new-generation-artificial -intelligence-development-plan-2017.

Russia's deployment of the Uran-9 unmanned ground vehicle in Syria: Sebastien Roblin, "What Happened When Russia Tested Its Uran-9 Robot Tank in Syria?," *National Interest*, April 7, 2021, https://nationalinterest .org/blog/reboot/what-happened-when-russia-tested-its-uran-9-robot -tank-syria-182143.

US DoD's Artificial Intelligence Strategy: "Fact Sheet: 2018 DoD Artificial Intelligence Strategy: Harnessing AI to Advance Our Security and Prosperity," US Department of Defense, February 12, 2019, https://media .defense.gov/2019/Feb/12/2002088964/-1/-1/1/DOD-AI-STRATEGY -FACT-SHEET.PDF.

Hicks's speech on AI: Kathleen Hicks, "The State of AI in the Department of Defense," US Department of Defense, November 2, 2023, www.defense .gov/News/Speeches/Speech/Article/3578046.

Drones in Ukraine war: Amir Husain, "2022: A Look Ahead," *Forbes.com*, January 3, 2022, www.forbes.com/sites/amirhusain/2021/12/31/2022-a -look-ahead.

Drone-to-drone combat: "Drone Wars: Russian FPV Takes Down Ukrainian 'Baba Yaga' as Machines Fight Each-Other," *Daily Mail*, April 15, 2024, www.youtube.com/watch?v=KmlVSALESeE.

Russian and Ukrainian drone deployments: Mariano Zafra, Max Hunder, Anurag Rao, and Sudev Kiyada, "How Drone Combat in Ukraine Is Changing Combat," Reuters, March 26, 2024, www.reuters.com/graphics /UKRAINE-CRISIS/DRONES/dwpkeyjwkpm.

Russia's use of the Orion UAV and Uran-9 UGV in the Ukraine conflict: "Russia's Increasing Use of Unmanned Ground Vehicles in Ukraine Conflict," Army Recognition Group, March 31, 2024, https: //armyrecognition.com/focus-analysis-conflicts/army/conflicts-in-the -world/russia-ukraine-war-2022/russia-s-increasing-use-of-unmanned -ground-vehicles-in-ukraine-conflict.

Israel's war with Hamas: Marwa Fatafta and Daniel Leufer, "Artificial Genocidal Intelligence: How Israel Is Automating Human Rights Abuses and War Crimes," accessnow.org, May 9, 2024, www.accessnow.org/publication /artificial-genocidal-intelligence-israel-gaza; "AI-Assisted Targeting in the Gaza Strip," Wikipedia, accessed September 28, 2024, https://en .wikipedia.org/wiki/AI-assisted_targeting_in_the_Gaza_Strip; Robert A. Pape, "Hamas Is Winning: Why Israel's Failing Strategy Makes Its Enemy Stronger," *Foreign Affairs*, June 21, 2024, www.foreignaffairs.com/israel /middle-east-robert-pape; Emanuel Fabian, "Drone Strike Hits Gunmen Trying to Loot Gaza Aid Convoy," *Times of Israel*, September 21, 2024.

Orbiter 1K: "Aeronautics Signs USD 40 Million Deal for Orbiter 1K Loitering Munitions," Defense Industry Europe, August 31, 2023, https: //defence-industry.eu/aeronautics-signs-usd-40-million-deal-for -orbiter-1k-loitering-munitions.

Guardium: Andrew Tarantola, "This Unmanned Patroller Guards Israeli Borders for Days on End," *Gizmodo*, September 14, 2012, https://gizmodo .com/this-unmanned-patroller-guards-israelis-borders-for-days-5943055.

Hamas's cybernetic weapons: Kerry Chávez, "How Hamas Innovated with Drones to Operate Like an Army," *Bulletin of the Atomic Scientists*, November 1, 2023, https://thebulletin.org/2023/11/how-hamas-innovated -with-drones-to-operate-like-an-army.

Challenges facing robots: Bernd Debusmann Jr., "Can Football-Playing Robots Beat the World Cup Winners by 2050?," *BBC News*, September 26, 2021, www.bbc.com/news/business-58662246.

Building-Wide Intelligence project: www.cs.utexas.edu/~larg/bwi_web.

Hyperwar: Amir Husain et al., *Hyperwar* (SparkCognition Press, 2018).

Air combat and drones: Paul Iddon, "Iran and Turkey Are Betting on Drone Aircraft Carriers to Project Power," *Business Insider*, September 7, 2024, www.businessinsider.com/iran-turkey-drone-carrier-ships-project -power-2024-9.

Exoskeletons: Jared Keller, "The Inside Story Behind the Pentagon's Ill-Fated Quest for a Real-Life 'Iron Man' Suit," *Task and Purpose*, July 11, 2021, https://taskandpurpose.com/news/pentagon-powered-armor-iron-man -suit.

CHAPTER 7: CLIODYNAMICS AND CYBERNETICS

Broad overview of Peter Turchin's work: "Cliodynamics: History as Science," https://peterturchin.com/cliodynamics-history-as-science.

Seshat: https://seshatdatabank.info.

Turchin's concept of "secular cycles" in historical societies: Peter Turchin and Sergey Nefedov, *Secular Cycles* (Princeton University Press, 2009).

Turchin's "demographic-structural theory": Peter Turchin, *Historical Dynamics* (Princeton University Press, 2003); see also https://peterturchin.com/structural-demographic-theory.

Elite overproduction and popular immiseration in the United States: Peter Turchin, *End Times* (Penguin, 2023).

Concept of "skill-biased technological change" favoring skilled workers over unskilled workers: Eli Berman, John Bound, and Stephen Machin, "Implications of Skill-Biased Technological Change: International Evidence," *Quarterly Journal of Economics* 113, no. 4 (November 1998): 1245–1279, www.jstor.org/stable/2586980; see also www.studysmarter.co.uk/explanations/macroeconomics/international-economics/skill-biased-technological-change.

Criticism: David Card and John E. DiNardo, "Skill-Biased Technological Change and Rising Wage Inequality: Some Problems and Puzzles," *Journal of Labor Economics* 20, no. 4 (2002), https://davidcard.berkeley.edu/papers/skill-tech-change.pdf.

Potential impact of technological advancements on income inequality and elite overproduction: United Nations Economic and Social Commission for Asia and the Pacific, "Inequality in Asia and the Pacific in the Era of the 2030 Agenda for Sustainable Development: Chapter 4, Technology and Inequalities," 2018, www.unescap.org/sites/default/files/06Chapter4.pdf.

CHAPTER 8: WE ARE ALREADY OPTED IN

Gait analysis: Elsa J. Harris et al., "A Survey of Human Gait-Based Artificial Intelligence Applications," *Frontiers in Robotics and AI*, January 2, 2022, www.frontiersin.org/journals/robotics-and-ai/articles/10.3389/frobt.2021.749274/full.

Mission: Impossible: David Viramontes, "Every 'Mission: Impossible' Mask Reveal, Ranked," *Variety*, July 31, 2020, https://variety.com/lists/mission-impossible-mask-reveals-ranked/living-manifestation-of-destiny-mission-impossible-rogue-nation.

Data on apps and their use: "Mobile App Download Statistics & Usage Statistics (2024)," Buildfire, accessed September 28, 2024, https://buildfire.com/app-statistics.

Associated Press investigation revealing Google's location-data storage practices: Ryan Nakashima, "AP Exclusive: Google Tracks Your Movements, Like It or Not," Associated Press, August 13, 2018, https://apnews.com/article/828aefab64d4411bac257a07c1af0ecb.

Google ad revenue: Tiago Bianchi, "Advertising Revenue of Google from 2001 to 2023," *Statista*, May 22, 2024, www.statista.com/statistics/266249/advertising-revenue-of-google.

Microsoft's Azure revenue in the first quarter of 2021: Akanksha Rana and Julia Love, "Microsoft Sees Steady Cloud Growth After Record Quarterly Profit," Reuters, July 27, 2021, www.reuters.com/technology/microsoft-beats-quarterly-revenue-estimates-2021-07-27.

Waze's collection of real-time location and movement data: Ryan Daws, "Ru Roberts, Waze: On Improving Journeys Using Real-Time Data," *IoT News*, February 17, 2022, https://iottechnews.com/news/ru-roberts-waze-on-improving-journeys-using-real-time-data.

General Data Protection Regulation: https://gdpr-info.eu.

CHAPTER 9: THE TECHNOLOGIES OF FREEDOM

Sir Tim Berners-Lee's Solid project for decentralized personal data stores: https://solidproject.org.

Inrupt's development of personal data stores: Philippe Haenebalcke, "What Do Personal Data Stores Mean for Privacy Regulations, Digital Identity, and Customer Trust?," Inrupt, February 28, 2023, www.inrupt.com/blog/impacts-of-personal-data-stores-on-regulations-and-customers.

NYC Mesh's community-owned mesh network in New York City: www.nycmesh.net.

National Institute of Standards and Technology's (NIST) call for post-quantum cryptographic algorithms: https://csrc.nist.gov/projects/post-quantum-cryptography. See also Chad Boutin, "NIST Announces First Four Quantum-Resistant Cryptographic Algorithms," NIST, July 5, 2022, www.nist.gov/news-events/news/2022/07/nist-announces-first-four-quantum-resistant-cryptographic-algorithms.

IBM's integration of CRYSTALS-Kyber and CRYSTALS-Dilithium in its z16 system: Anne Dames, "Available on IBM z16: Future-Proof Digital Signatures with a Quantum-Safe Algorithm Selected by NIST," IBM, July 26, 2022.

Cloudflare's Interoperable, Reusable Cryptographic Library (CIRCL), including Kyber: Kris Kwiatkowski and Armando Faz-Hernández, "Introducing

CIRCL: An Advanced Cryptographic Library," Cloudflare, June 20, 2019, https://blog.cloudflare.com/introducing-circl.

Amazon's AWS: "AWS Secrets Manager Connections Now Support the Latest Hybrid Post-Quantum TLS with Kyber," Amazon Web Services, August 2, 2022, https://aws.amazon.com/about-aws/whats-new/2022/08/aws-secrets-manager-connections-support-hybrid-post-quantum-tls-kyber.

EnerPort in Ireland and Brooklyn Microgrid as examples of peer-to-peer energy-trading platforms: "EnerPort: Irish Blockchain Project for Peer-to-Peer Energy Trading," *Energy Informatics* 1, article no. 14, 2018, https://energyinformatics.springeropen.com/articles/10.1186/s42162-018-0057-8; see also "Brooklyn Microgrid News," www.brooklyn.energy/press.

CHAPTER 10: CYBERNETIC SYNTHESIS

Facebook's emotional contagion experiment: Adam D. I. Kramer, Jamie E. Guillory, and Jeffrey T. Hancock, "Experimental Evidence of Massive-Scale Emotional Contagion Through Social Networks," *Proceedings of the National Academy of Science* 111, no. 24, www.pnas.org/doi/10.1073/pnas.1320040111.

InterPlanetary File System (IPFS) and Filecoin as decentralized storage solutions: https://ipfs.io and https://filecoin.io.

CNAS report on strategic competition: Michael Horowitz et al., "Strategic Competition in an Era of Artificial Intelligence," CNAS, July 25, 2018, www.cnas.org/publications/reports/strategic-competition-in-an-era-of-artificial-intelligence.

Study on negative mood increasing support for conservative political views: Kyle Nash and Josh Leota, "Political Orientation as Psychological Defense or Basic Disposition? A Social Neuroscience Examination," *Cognitive, Affective, & Behavioral Neuroscience* 22 (2022): 586–599, https://link.springer.com/article/10.3758/s13415-021-00965-y.

Study on sad mood: Nitika Garg and Jennifer S. Lerner, "Sadness and Consumption," *Journal of Consumer Psychology* 23, no. 1 (January 2013), www.sciencedirect.com/science/article/abs/pii/S1057740812000757.

Study on "inevitable" unpleasant outcomes: Lyn Y. Abramson et al., "Learned Helplessness in Humans," *Journal of Abnormal Psychology* 87, no. 1 (1978): 49–74.

New York Times investigation: Jennifer Valentino-DeVries et al., "Your Apps Know Where You Were Last Night, and They're Not Keeping It Secret,"

New York Times, December 10, 2018, www.nytimes.com/interactive /2018/12/10/business/location-data-privacy-apps.html.

Wall Street Journal on Google tracking: John D. McKinnon, "Google Reaches $391.5 Million Settlement with States over Location Tracking Practices," *Wall Street Journal*, November 14, 2022, www.wsj.com/articles /google-reaches-391-5-million-settlement-with-states-over-location -tracking-practices-11668444749.

Facebook's "shadow profiles": Andrew Quodling, "Shadow Profiles—Facebook Knows About You, Even if You're Not on Facebook," *Conversation*, April 13, 2018, https://theconversation.com/shadow-profiles-facebook -knows-about-you-even-if-youre-not-on-facebook-94804.

Cambridge Analytica scandal: "Cambridge Analytica: Facebook Row Firm Boss Suspended," *BBC News*, March 20, 2018, www.bbc.com/news/uk -43480048.

INDEX